MW00427413

RECYCLING
CONSTRUCTION &
DEMOLITION WASTE

McGRAW-HILL'S GREENSOURCE SERIES

Attmann
Green Architecture: Advanced Technologies and Materials

Gevorkian
Alternative Energy Systems in Building Design
Solar Power in Building Design: The Engineer's Complete Design Resource

GreenSource: The Magazine of Sustainable Design
Emerald Architecture: Case Studies in Green Building

Haselbach
The Engineering Guide to LEED—New Construction: Sustainable Construction for Engineers

Luckett
Green Roof Construction and Maintenance

Melaver and Mueller (eds.)
The Green Building Bottom Line: The Real Cost of Sustainable Building

Nichols and Laros
Inside the Civano Project: A Case Study of Large-Scale Sustainable Neighborhood Development

Winkler
Recycling Construction & Demolition Waste: A LEED-Based Toolkit

Yudelson
Green Building Through Integrated Design
Greening Existing Buildings

About *GreenSource*
A mainstay in the green building market since 2006, *GreenSource* magazine and GreenSourceMag.com are produced by the editors of McGraw-Hill Construction, in partnership with editors at BuildingGreen, Inc., with support from the U. S. Green Building Council. *GreenSource* has received numerous awards, including American Business Media's 2008 Neal Award for "Best Website" and 2007 Neal Award for "Best Start-up Publication," and FOLIO magazine's 2007 Ozzie Awards for "Best Design, New Magazine" and "Best Overall Design." Recognized for responding to the needs and demands of the profession, *GreenSource* is a leader in covering noteworthy trends in sustainable design and best practice case studies. Its award-winning content will continue to benefit key specifiers and buyers in the green design and construction industry through the books in the *GreenSource* Series.

About McGraw-Hill Construction
McGraw-Hill Construction, part of The McGraw-Hill Companies (NYSE: MHP), connects people, projects, and products across the design and construction industry. Backed by the power of Dodge, Sweets, *Engineering News-Record (ENR)*, *Architectural Record*, *GreenSource*, *Constructor*, and regional publications, the company provides information, intelligence, tools, applications, and resources to help customers grow their businesses. McGraw-Hill Construction serves more than 1,000,000 customers within the $4.6 trillion global construction community. For more information, visit www.construction.com.

About the International Code Council
The International Code Council (ICC), a membership association dedicated to building safety, fire prevention, and energy efficiency, develops the codes and standards used to construct residential and commercial buildings, including homes and schools. The mission of ICC is to provide the highest quality codes, standards, products, and services for all concerned with the safety and performance of the built environment. Most U.S. cities, counties, and states choose the International Codes, building safety codes developed by the ICC. The International Codes also serve as the basis for construction of federal properties around the world, and as a reference for many nations outside the United States. The ICC is also dedicated to innovation and sustainability and Code Council subsidiary, ICC Evaluation Service, issues Evaluation Reports for innovative products and reports of Sustainable Attributes Verification and Evaluation (SAVE).

Headquarters: 500 New Jersey Avenue, NW, 6th Floor, Washington, DC 20001-2070
District Offices: Birmingham, AL; Chicago, IL; Los Angeles, CA
1-888-422-7233
www.iccsafe.org

RECYCLING
CONSTRUCTION &
DEMOLITION WASTE
A LEED-BASED TOOLKIT

GREG WINKLER

New York   Chicago   San Francisco   Lisbon   London   Madrid
Mexico City   Milan   New Delhi   San Juan   Seoul
Singapore   Sydney   Toronto

The McGraw·Hill Companies

Cataloging-in-Publication Data is on file with the Library of Congress.

McGraw-Hill books are available at special quantity discounts to use as premiums and sales promotions, or for use in corporate training programs. To contact a representative please e-mail us at bulksales@mcgraw-hill.com.

Recycling Construction & Demolition Waste: A LEED-Based Toolkit

1 2 3 4 5 6 7 8 9 0   DOC/DOC   1 9 8 7 6 5 4 3 2 1 0

ISBN 978-0-07-171338-2
MHID 0-07-171338-7

The pages within this book were printed on acid-free paper containing 100% postconsumer fiber.

Sponsoring Editor
Joy Bramble

Acquisitions Coordinator
Michael Mulcahy

Editorial Supervisor
David E. Fogarty

Project Manager
Kritika Kaul, Glyph International

Copy Editor
Upendra Prasad, Glyph International

Proofreader
Medha Joshi, Glyph International

Production Supervisor
Pamela A. Pelton

Composition
Glyph International

Art Director, Cover
Jeff Weeks

To my son Alexander
Keep building castles in the air
A few will find their way to earth

About the Author

Greg Winkler, AIA, LEED AP, has over 27 years of experience as a practicing architect and owner's representative on commercial, industrial, and institutional projects. A graduate of the Georgia Institute of Technology, Mr. Winkler has participated in construction projects as a design professional, design-builder, and owner's agent. This perspective gives him a pragmatic view of the possibilities—and pitfalls—of construction waste management.

Mr. Winkler currently works as the director of a regional construction trade organization. He is the co-author of McGraw-Hill's *Construction Administration for Architects*, a desktop guide for design professionals during construction.

CONTENTS

FOREWORD

Resource conservation and recycling have long been devoted passions for a segment of the population. As a vocation, recycling has been a cost-effective standard practice for centuries in the production of metals and it also has deep roots in the papermaking industry.

Both of these influences—the passion of environmental advocates and the cost-effectiveness sought in the business world—came together in the late twentieth century in the construction and demolition sectors.

With the exception of scrap metals, most other materials generated at a construction or demolition site had previously been regarded as waste—as garbage or rubble. The procedure that developed (and it developed as an efficient one) was to divert materials by the truckload to inert landfills designed to accept this rubble or waste.

Construction scrap and demolition debris was largely out of sight and out of mind for conservationists who focused instead on the pop bottles, newspapers, and other unrecycled litter that was clearly visible.

But this sensibility began to change when waste characterization studies (essentially landfill audits) began to be conducted. Statistics revealed quite clearly that the scrap and debris being generated at construction sites added up to a considerable volume—and environmental advocates noticed.

On the cost-effectiveness side, contractors also began to see changes in the factors that affect material hauling and disposal costs. Sprawling metropolitan areas often meant landfills were a longer journey away—just as diesel fuel costs were rising. Similarly, stone being brought in from quarries was traveling longer distances and triggering higher transportation costs.

The concept and practice of crushing and recycling concrete and asphalt began to grow rapidly, and in many cases spurred entrepreneurs to try similar approaches with other construction and demolition (C&D) materials, such as wood scrap and gypsum drywall scrap.

The two universes—contractors finding cost-effective ways to recycle and advocates identifying C&D materials as a major landfill contributor—were soon joined by a third group of interested parties: architects, builders, and property owners who wanted to demonstrate environmental responsibility.

These construction industry professionals spurred the rapid rise of the green building movement, including the U.S. Green Building Council and its LEED (Leadership in Energy and Environmental Design) standards. Turning C&D materials into recyclable commodities has been adopted as part of the LEED standards, providing a boost to an emerging economic sector that had already gained considerable momentum.

Labeling change as revolutionary can run the risk of overstating a case. In many ways, however, the manner in which C&D scrap materials are now handled at a job site—whether diverted into a recycling container or processed for recycling on site—has changed in ways that would be difficult to recognize by someone who had not set foot on a construction site since the 1980s.

BRIAN TAYLOR
Editor-in-Chief
Construction & Demolition Recycling magazine
www.CDRecycler.com
www.RecyclingToday.com

PREFACE

In the late 1980s the National Trust for Historic Preservation (U.S.) embraced a concept called *embodied energy*. Spawned from the oil embargo of the late 70s, this concept was based on the assertion that the accumulated labor, transport, materials manufacturing, and erection of our buildings represented captured energy, and that preserving and reusing existing buildings was among the most effective forms of energy conservation. The concept never gained ground within or outside of preservation circles, principally because the best reasons for preserving buildings—their architectural or historic merit—are far more compelling than that of not wasting energy by demolishing them.

In some ways, the original arguments for recycling construction and demolition waste ran along the same fault lines. Proponents argued, with good cause, that recycling C&D waste was the right thing to do for society and the environment. The moral argument was undeniable, but in the early days of C&D recycling few markets existed outside of scrap metal for the materials generated in the demolition of a building. Absent the economic impetus, the social arguments for recycling were simply not enough to compel change in our long-standing "throw-away" culture. Contracting is a competitive business, and until the landfills and incinerators of America became too expensive and politically unpalatable for communities to tolerate, they were the preferred and least costly way of disposing of our nation's large stream of C&D waste. In this sense, the demolition and contracting industries were a reflection of society at large. Thankfully, society and the contracting profession have changed. Recycling is now the preferred, and often required, means of handling waste from construction sites.

This happened because the public became more accepting of purchasing manufactured goods with recycled content. Manufacturers recognized this acceptance, as well as the public goodwill that using recycling content in their products generated, and developed more innovative ways to use waste material. It also happened because of a gradual shift in the economics of waste management. Contractors learned how to better manage, sort, and process material on their jobsites to make it more marketable. Recyclers became more sophisticated in separating single-stream materials and in processing waste to manufacturer's specifications. Manufacturers learned that recycled content yielded economic and marketing benefits, and consumers realized they could buy recycled-content products with confidence. Recycling markets for C&D waste have grown dramatically in the past decade, and continue to expand even in economic hard times.

For that reason the market sections of this book, or any book on recycling waste, are a snapshot in time of a continuously changing landscape. Similarly, the code and certification requirements associated with recycling are in a constant state of evolution. The management techniques and tools—the means of managing waste on a construction site—are longer lasting. Even they will evolve, however, as the opportunities and markets for recycling C&D waste expand.

All of this argues for the demolition and contracting professions to fully commit to learning and maintaining awareness of the best current practices in recycling their jobsite waste. There is no longer any excuse for a contractor to not recycle at least half the waste generated from his or her construction site. In urban areas with better developed markets, contractors should be able to easily recycle 75 percent of their waste. Those who recycle more, and do it better, will be more profitable and competitive.

In the realm of managing construction and demolition waste, social accountability and financial accountability have finally merged on the side of recycling. It is a good thing.

GREG WINKLER

ACKNOWLEDGMENTS

Gratitude to Rob Nigro for his skillful editing and friendship.

Appreciation to Hamid Naderi, P.E. of the International Code Conference (ICC) for his review of the manuscript and his organization's support of the book.

Thanks to Brian Taylor, editor of *Construction & Demolition Recycling* magazine, for writing the Foreword.

Thanks to Kritika Kaul and the professionals at Glyph International for their work in bringing the book to publication.

Deep appreciation to Joy Bramble, Senior Editor at McGraw-Hill Professional for her proposal guidance and support.

Special gratitude to my wife Lisa for her love and tolerance, but especially for her English to SI conversion skills!

Respect and admiration for the construction and demolition professionals who skillfully manage recycling programs on construction sites across the country.

RECYCLING CONSTRUCTION & DEMOLITION WASTE

RECYCLING WASTE:

THE FUNDAMENTALS

The amount of waste generated by building construction and demolition activities in the United States is enormous—more than 164 million tons (149 million metric tonnes) per year. This represents 25 to 40 percent of all the discarded solid waste in the country. What happens to all of this material? More than 75 percent is trucked to landfills or incinerators.[1] Increasingly, municipalities and building owners are responding to the cost of maintaining landfills and the burgeoning environmental movement in the country by requiring that contractors recycle construction and demolition waste. The U.S. Green Building Council (USGBC) estimates that as much as 95 percent of the waste on a typical construction site can be recycled. In response, green waste management ordinances are appearing in municipalities across the nation.

Savvy developers and owners are also realizing the cost savings of recycling versus disposal, and are requiring their design professionals to incorporate recycling waste requirements into their construction documents.

The life cycle of a building used to be a one-way street. Building materials were extracted and used to manufacture building products, and once the building reached the end of its useful life and was demolished, the materials were buried in a landfill or incinerated. Societal and economic factors require that today's building life cycle be circular (see Fig. 1.1), with the loop completed to the largest extent possible by reusing demolition materials to manufacture new products.

The Reasons to Recycle

The original reasons for the existence of landfills were simplicity and economy. It was easier and less expensive to send demolished building materials and construction waste to landfills than to attempt to recycle them. Indeed, the recycling markets barely existed for many demolition materials even 10 years ago. Contractors seeking

Figure 1.1   The building products life cycle.

markets for any waste beyond asphalt, brick, or concrete would find few takers, as manufacturers had not developed a wide range of products using recycled waste, and still often preferred to use virgin materials to control quality and costs. However, as raw materials became scarcer and more expensive to extract and municipalities began resisting landfill expansions, the economics of waste management flipped in favor of recycling. Manufacturers began to look at recycled waste as a more reliable and cost-effective supply source for raw material, and altered existing products or developed new ones to better use recycled products. At the same time, these companies also began to realize the marketing and public relations benefits of touting their new-found philosophy of earth stewardship. The economic case for recycling has become a solid argument for contractors to adopt a hierarchal approach to waste management on their sites (see Fig. 1.2).

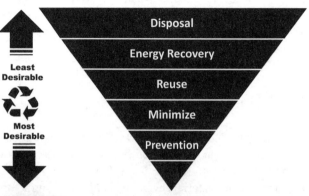

Figure 1.2   The recycling triangle.

Here are the top reasons why recycling is a smarter, more profitable, and more sustainable way to deal with construction and demolition waste:[2]

1 *Employment*: Ecocycle.org estimates that for each job in a landfill, 10 other people are employed elsewhere in processing recycled products and another 25 are employed in manufacturing products from recycled materials.

2 *Costs*: Landfills and incinerators are economic disasters. Roughly 20 percent of the Superfund sites on the U.S. Environmental Protection Agency (EPA) list are landfills, and all landfills require long-term monitoring to check for toxic *leachate* (also known as "garbage juice"). Incinerators require large capital investments and require a continual stream of garbage to remain economical. Even the most efficient incinerators emit dioxin, mercury, lead, and cadmium into the environment. The fact that no new incinerators have been constructed in the United States since 1995 is powerful evidence that they are not economical alternatives to recycling.

3 *Energy*: Recycling saves energy through reducing the net amount of energy expended in extracting and using raw materials. For example, for every one million tons of aluminum material recycled, Americans save the equivalent of 35 million barrels (5.6 million m^3) of oil. For every million tons of PET or HDPE plastic recycled, U.S. consumers save approximately 9 million barrels (1.4 million m^3) of oil.

StopWaste.org estimates that commercial construction projects in the United States generate between 2 and 6 lb of solid waste per square foot of building area. Residential construction generates even higher levels: 3 to 15 lb/ft^2 (14 to 73 kg/m^2). Using a median of 8.5 lb (3.8 kg), a standard 2000-ft^2 (186-m^2) home will create 17,000 lb (7711 kg) of construction waste—more than 8 tons (7.3 metric tonnes).[3]

On a larger scale, a 2007 survey by the Construction Materials Recycling Association (CMRA) estimated that more than 140 million tons (127 million metric tonnes) of concrete were being recycled in the United States that year, making concrete the most recycled material (by weight). The total construction waste stream, as estimated by CMRA, was approximately 325 million tons (297 million metric tonnes) in 2007 (including work on roads and bridges).[4]

RECYCLING CHALLENGES AND OPPORTUNITIES

Despite the dramatic growth in construction and demolition (C&D) waste recycling, the EPA estimated in 2003 that only 20 percent of the waste generated on an average construction site was being recycled or reused. Also according to the EPA, demolition accounts for 53 percent of construction industry waste, renovation 38 percent, and new construction 9 percent.[5] Clearly then, the greatest opportunity to increase jobsite recycling rates would appear to be in the demolition field. Since this realm of construction is totally under the control of the contractor, it would appear equally clear that this is an area where the demolition and construction industry should seize the moment.

Not so fast, many contractors would argue. Despite contractors' best efforts to recycle, some products either do not have markets at all, or have recycling opportunities that are limited to select areas of the country. Furthermore, the construction industry is driven more than any other by small economies. Absent other mandates, the money saved by landfilling may be the way a contractor is able to bid low enough to win a project or perform it well enough to make money on it. Recycling works in the construction industry when everyone is required to use it, or where the markets exist to make it less costly than landfilling. Despite significant growth in the recycling industry in the past 10 years, more recycling markets are needed to enable contractors to consistently market their C&D waste.

To realize the difficulties some contractors face in finding markets for recycled materials, consider the case of carpet recycling in the United States. In 2002, the Carpet and Rug Institute of America, which represents the country's largest carpet manufacturers, signed an agreement with the EPA to recycle 40 percent of all used carpet products by the year 2012. The Institute's Carpet American Recovery Effort (CARE) program set as a goal the recycling of 2 billion lb (907 million kg) of carpet each year.[6] To fulfill this ambitious goal, the institute had developed specific plans for how and where to recycle their products. Along the way, however, the Institute learned a few things.

Their original plan was to sell the bulk of the recycled product to cement kilns as fuel, a purpose for which it was well-suited. Cement manufacturers, however, were only willing to deal with the extra trouble of using carpet waste as fuel if the carpet manufacturers *paid* them to take it. Other efforts to use the waste to make new consumer products, such as auto parts, were successful to varying extents. But the small quantities of products and the unreliability of the markets could only soak up a limited amount of the available waste. It turns out the best use of waste carpet products is in making new carpet, and in this goal the carpet industry has been reasonably successful. Still, as of 2006, the industry was only successful in diverting 253 million lb (115 million kg) of used carpeting from landfills, a little more than 10 percent of their original goal. Despite falling well short of their goal, the carpet industry's recycling rate is still far better than that achieved by other industries, such as those producing gypsum, resilient flooring, and ceiling tiles. Can this problem be cured by state or national legislation requiring manufacturers to begin using more recycled materials in their new products, thereby creating more markets? Not necessarily. As Robert Cassidy of *Building Design & Construction* magazine notes: "California now requires 10 percent postconsumer content in carpet. As a result, some producers are mixing glass into their carpet backing to reach 10 percent recycled content. Twenty years from now, that glass-backed carpet will be nearly impossible to recycle."[7]

The simple reality is that recycling jobsite waste is not solely the responsibility of the contractor, and building material manufacturers and architects must contribute to creating spaces that are more *recyclable*.

The Waste Management Streams

C&D waste flowing from a construction site follows one of three typical paths: landfill, single-stream, or source separated. These paths may vary by project, of course, and a single project may include all three methods as contractors match waste to local markets, and deal with hazardous waste in whatever manner is best suited to the area regulations.

The traditional method of disposing of C&D waste was to take it straight to a land-fill. Though still allowed in some areas of the country, municipalities are increasingly demanding that contractors document at least an attempt to recycle waste products from their site. With the increasing use of crushed concrete as road aggregate and wood waste for a number of uses, there are very few areas of the country where some market for recycled materials does not exist. Even municipalities who have not yet adopted a threshold recycling rate for demolition projects are requiring contractors to file a plan showing how they will manage the project waste. With existing landfills increasingly reaching capacity and being recognized as long-term hazards, and the reluctance of communities to allow the construction of new facilities, recycling has become the only means of disposing of materials in some urban markets. Between 1996 and 2002, the *Whole Building Design Guide* estimates 26 percent of C&D waste disposal landfills closed in the United States.[8] The closing of the Fresh Kills landfill in New York City in 1996, for instance, removed the last remaining landfill from the waste disposal market for that metropolitan area. The result was a dramatic increase in the cost of tipping fees at city transfer stations as exporting waste via truck became the only option for dispos-ing of waste. Recycling markets for C&D waste mushroomed as the economics tipped in favor of reusing instead of disposing.

Over the past decade this same scenario has played out in similar, though less dra-matic fashion, in urban areas across the country. Although the effect of the LEED (Leadership in Energy and Environmental Design) program in creating a public demand for recycling cannot be dismissed, the driving force in creating more recycling markets was the simple shift in economics that made recycling and reusing less expensive than landfilling.

As a result, two basic types of waste streams have developed to handle C&D waste. The most common is called source separation. In this method, contractors sort the materials on the jobsite into discrete containers according to what the market demands. This work is often carried out by demolition contractors or the subcontractors working on the site, and usually results in the highest value (or lowest cost) in recycling waste. This system also results in the least amount of waste ending up in a landfill, since the contractor is working with a number of different markets to provide the specific products they can process for end users.

The other method, less common on construction sites, is single-stream, or com-mingled waste recycling. In this system, wastes of various types are collected on the jobsite in common containers and transported to a recycler who separates them into marketable components at his facility. This system is becoming more widely used in municipal household collection because it results in higher collection rates of recy-clables than does a source-separated system, in which homeowners are asked to separate metal, glass, plastic, and paper. With commingling, some of the same collection rate increases occur on a construction site, since the ease of compliance and more limited separation rules make it simpler for subcontractors to recycle. However, two facts currently work against using single-stream recycling extensively for C&D waste: it

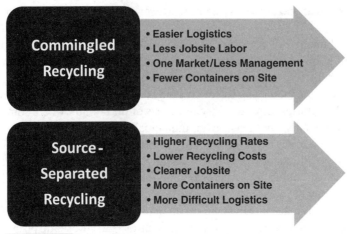

Figure 1.3   Commingled versus source-separated recycling.

saves less money than source separation, and the contractor can use subcontractors to perform the separation work. See Fig. 1.3 for a summary of the relative pros and cons of the two systems.

As single-stream sorting facilities become more widespread and cost-effective, this reality may shift. But for now, contractors will normally find that sorting waste at the construction site is the best way to minimize their C&D waste costs. Once the materials are sorted, various markets open up for the reuse of waste from a construction site. Here are some of the more common uses for typical C&D waste:

1 *Metal*: Almost all the metal waste from a project goes to scrap dealers for feedstock in creating new products. This is true for metals ranging from flashing to steel W sections.

2 *Concrete*: The market for recycled concrete aggregate is dependent on demand and distance. In major cities where demand is high, landfill costs are high, and transportation distances are short, it is easy to market concrete waste. In rural areas with lower landfill costs and longer distances from the mill to the market, the economics of recycling may not yet be competitive with disposal.

3 *Wood*: Areas with wood-burning power plants or a manufacturing base that uses scrap wood as a fuel source provide a ready market for wood waste. In this market, demolition waste competes with cleaner virgin wood waste generated by manufacturing operations. While both are forms of recycling, the cleaner fuel will be more desirable than wood waste with metal and paint contaminants.

There is a surprising range of markets for a wide variety of demolition waste. That these markets are not always extensive or widespread does not detract from the fact that our economy has proven that it can effectively reuse many kinds of C&D waste, including:

- Landscaping and debris resulting from land clearing (green wood materials)
- Asphalt pavement
- Gravel and aggregate products
- Concrete
- Masonry scrap and rubble (brick, concrete masonry, stone)
- Metals (ferrous and nonferrous)
- Clean wood (dimensional lumber, sheet goods, millwork, scrap, pallets)
- Plastics (films, containers, PVC products, polyethylene products)
- Asphalt/bituminous roofing
- Insulation materials
- Glass (untempered)
- Door and window assemblies
- Carpet and carpet padding
- Fibrous acoustic materials
- Ceiling tiles
- Plumbing fixtures and equipment
- Mechanical equipment
- Lighting fixtures and electrical components
- Cardboard packing and packaging
- Other waste items

In assessing how recycled materials are classified in new products, it is important to understand a couple of basic terms common in the recycling industry:

- *Postconsumer recycled content* indicates that materials have been purchased once and have already been used by consumers. Products with a high percentage of post-consumer recycled content represent a very efficient use of our resources.
- *Postindustrial recycled content* indicates that manufacturing waste has been cycled back into the production process. These products do not represent the significant resource savings that postconsumer products do, but are usually preferable to those that use solely virgin materials.

The Economic Case

In the study *Recycling Construction & Demolition Wastes*, prepared by the Institution Recycling Network, author Mark Lennon makes a compelling argument for the economic benefit of recycling. In the Boston area, the cost of landfill disposal of mixed construction debris (concrete, brick, and block) ran approximately $136 per ton in 2005, including $31 per ton of transportation costs. The cost to recycle this mixed debris, including transportation, was $21 per ton—saving 84 percent over disposal costs.[9] In the worst case, argues Lennon, recycling saves at least 50 percent of the cost of landfilling on practically any waste component of a project. On major demolition projects, with considerable amounts of solid waste, the savings from recycling versus disposal can be significant, and can literally turn a break-even project into a profit maker. Even on smaller projects, recycling savings can be significant.

All About LEED® and Other Certifications

LEED® stands for *Leadership in Energy and Environmental Design*, and is the registered trademark of a program developed by the USGBC to promote sustainable practices in building construction and renovation. LEED is a points-based system in which building projects earn points by satisfying specific green building criteria defined by USGBC in seven credit categories. The six categories are: (1) sustainable sites, (2) water efficiency, (3) energy and atmosphere, (4) materials and resources, (5) indoor environmental quality, and (6) innovation in design.

First developed in 1998, LEED has grown to include standards for eight categories of projects, most notably new construction, existing buildings, and neighborhood development. To date, more than 35,000 LEED-certified projects in all 50 states and 91 countries have been completed or are in the planning process, representing more than 4.5 billion ft² (418 million m²) of building area.[10]

LEED encourages recycling of construction and demolition waste through its credits on Materials and Resources (see Chap. 7: Compliance Connection). These credits are granted for on-site waste diversion for two thresholds: in excess of 50 or 75 percent of the total solid waste generated on a project. More than any other single factor, the LEED program has raised public awareness of the large amount of waste generated by construction activities and has driven owners and governments to work toward diverting this waste away from landfills and incinerators and into a wider market for recycled materials.

A host of other sustainable certification programs and design guides are available to promote the use of green products and encourage more energy-efficient design and construction. See Fig. 1.4.

These programs are discussed in depth in Chap. 8: Other Green Certification and Code Programs. Most notable are two programs with different intents: Green Globes program and ICC-ES SAVE™ program.

GREEN GLOBES PROGRAM

The *Green Globes Program* is a voluntary certification program administered by the Green Building Initiative, a broad-based consortium of industry, government, and nonprofit representatives who modified an early Canadian program into an online resource that is promoted as a more streamlined and interactive alternative to LEED.

Compliance Connection

- LEED (US Green Building Council)
- ICC-ES SAVE program (International Code Council)
- International Green Building Code (International Code Council)
- Green Building Program (National Association of Home Builders)
- Energy Star Program (US Environmental Protection Agency)
- Whole Building Design Guide (National Institute of Building Sciences)
- Green Globes Program (Green Building Initiatives)
- Green Home and Commercial Building Standards (Florida)
- Green Point Rated Program (California)
- Breeam (United Kingdom)
- Green Star (Australia)

Figure 1.4   **Prominent sustainable certification and design programs.**

Green Globe's focus on modeling energy-efficiency after occupancy is also favored over a sometimes burdensome commissioning system imposed by LEED requirements.

ICC-ES SAVE™ PROGRAM

Sustainable Attributes Verification and Evaluation™ (SAVE™) Program   The International Code Council (ICC) is the preeminent code authority in the country, dedicated to creating model codes that promote building safety, fire prevention, and energy efficiency for residential and commercial buildings. ICC codes are in use, to varying extents, in all 50 states. The International Code Council has created the ICC-ES SAVE™ program to verify manufacturers' claims regarding the sustainable characteristics of their products. One of the largest problems arising out of the popularity of sustainable construction is the practice of *greenwashing*, or making environmental claims for a product that cannot be documented or are patently false. The purpose of the SAVE program evaluation is to allow manufacturers the opportunity to voluntarily document the sustainable attributes of their products. Once approved by ICC, this documentation can help those seeking to qualify for points under major green rating systems LEED or Green Globes programs.

Energy-efficiency requirements have long been a staple of most national and international building codes. Municipalities, however, are increasingly weighing in with new provisions—and entire new codes—that address sustainable construction. The ICC is developing a set of green codes under a multiyear initiative called the *International Green Construction Code (IGCC: Safe and Sustainable by the Book)*. This intent of the initiative is to build a collaborative model code, using input from national organizations committed to sustainable construction, including the American Institute of Architects (AIA) and the American Society for Testing and Materials (ASTM), as well as outreach and feedback from the general public. (See Chap. 8: Other Green Certification and Code Programs for more information regarding this initiative.) Once completed, the ICC hopes that the International Green Construction Code will be widely adopted as a model sustainable construction code across the country.

The Choice of Method

Contractors now have a wider range of options in recycling construction and demolition waste. New markets, more competitive haulers, and more and more varied outlets for waste make it easier than ever to recycle a high percentage of C&D waste generated from a jobsite. More options, though, make it more imperative that the contractor research the best recycling path for his particular project on the project's front end. Early considerations in this evaluation include:

- Whether to use a single-stream method or source separation—or some combination of both?
- What demolition waste is marketable?
- What markets are available in the project area?
- How much site area is available for a recycling center?

The key question in the early analysis of how to approach waste recycling on a project is whether to use single-stream or source-separated methods. In some cases, the market in a particular area will dictate the choice, as single-stream recycling is not available as an option in many areas. In other cases, particularly urban areas, the contractor will have the choice of which method to use, and must consider the resulting financial and time options that accompany each choice. It is also likely that a contractor mapping out a recycling strategy for a project will find that a mix of options works for different situations. The recycling market in his area, for instance, might not pay a sufficient amount to warrant separating nonferrous metals, so why bother? In this case, the obvious solution is to sell the mixed metals to a recycler for the best price and save the on-site labor.

As with so many other decisions on a jobsite, it is possible to reap more financial benefits with more time investment. Since time is a finite resource in a construction project, and contractor staffing is expensive, the decision as to which recycling type to use carries major consequences. Following is a review of the two types of recycling, and their relative pros and cons. First, though, take a look at the old nonrecycling, landfill system in Fig. 1.5.

SINGLE-STREAM (COMMINGLED) RECYCLING

Single-stream, or commingled, recycling means placing all recyclable materials into a single container, which is then transported to a processing facility where different materials are separated by hand or by automated equipment. Contracting with a recycling

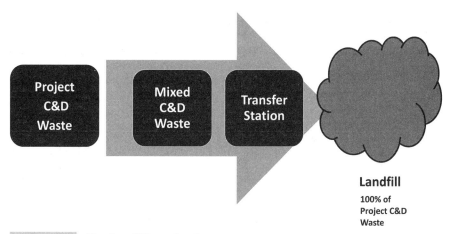

Figure 1.5   The landfill waste stream.

hauler who accepts commingled C&D materials can be advantageous, as it simplifies collection and increases the likelihood that any recycled material will be collected and deposited. See the steps involved in a single-stream system in Fig. 1.6.

Commingling services allow contractors to put multiple recyclables such as wood, cardboard, and metals in one container. The recycling company takes the materials to a sorting facility where the materials are separated for recycling. Materials not accepted in commingled loads will still need to be source-separated. Jobsite garbage must go in a separate container.

Commingled recycling simplifies the contractors' work in other ways as well. By dealing predominantly with one recycling hauler, they can reduce the administrative time and frustration of managing a number of different recycling pickups. The recycling

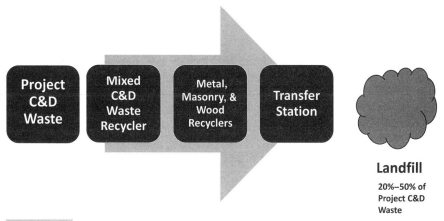

Figure 1.6   The commingled recycling waste stream.

area of the jobsite can be much smaller, as fewer (and larger) containers can be used for recycling waste. Subcontractor compliance rates are much higher for commingled recycling. The simplicity of the system results in higher participation and cleaner recycling areas.

There are also degrees of acceptable commingling and some constraints that recycling markets may place on what they accept. Certain recyclers may take commingled metals and concrete, for example, but may not accept wood in the mix. Others may only accept commingled materials that are free of drywall, insulation, or other waste that their system cannot easily sort from the mix. Despite the advantages and relative ease of use of commingled recycling, it results in some disadvantages for contractors in the field, including:

- *Higher costs*: Recyclers charge more to pick up commingled waste because they must absorb the costs of separating the waste later.
- *Less consistent loads*: Even though commingled recyclers can accept a wide range of waste in a single load, they cannot accept everything. The ease of commingling leads some workers to forget the limits, and adopt the mindset that practically any demolition material short of pure garbage can be recycled.
- *Recordkeeping difficulties*: Contractors using commingling must, out of necessity, chase various records well past their jobsite. While a contractor can certainly track the volume or weight of material being picked up and transported, he must rely on others to provide him with information about the breakdown of that load into various types of recyclables.
- *Compliance difficulties*: With commingled products, it can be more difficult for the field superintendent or project manager to determine if a particular subcontractor is fully complying with his recycling obligations. Since the waste must be going somewhere, it may be obvious in most situations that the subcontractor is contributing all his waste to the commingled collection container. On a busy demolition site, however, it may be difficult to make a rough visual assessment as to whether he is shortcutting the process through retaining valuable items (such as copper or other metals) to sell on his own, or diverting some recyclable waste to the garbage bin.

Where it is available, single-stream recycling offers contractors the advantages of low oversight, high participation rates, and savings over the cost of sending all waste to a landfill. It is an easy way to comply with municipal or owner requirements for diverting high percentages of waste from landfills. Single-stream recycling will not offer as many financial advantages as source separation, so the contractor is, in effect, paying the recycling company more to perform the separation for him. Since the main activity of a contractor is to build, he may find that using single-stream recycling frees his field superintendent and office staff from many of the management headaches of running a source-separation center on the jobsite.

On the other hand, on those projects where every dollar needs to be squeezed to ensure profitability, source separation offers more opportunities to save waste management costs.

SOURCE-SEPARATED RECYCLING

Source-separated recycling (also called source separation) is the alternative to commingling. The highest benefits of recycling come from separating waste materials at the jobsite and transporting and recycling them individually.

In this system, workers at the jobsite keep metals separate from wood, and wood separate from concrete, for example, and place each material into a different container. The containers are then transported by a recycler, transfer site, or directly to individual markets. Materials not accepted in commingled recycling, such as carpet or ceiling tile, must be source-separated for recycling (see Fig. 1.7 for a diagram of this recycling system).

Source separation has two major advantages over commingled recycling:

1 Source separation produces materials that are ready to go directly to market, saving the recycler or processor the cost of sorting materials.
2 Source-separated materials are generally of higher quality and have fewer contaminants, and are therefore worth more in recycling markets.

In less developed areas, source separation may be the only option for contractors to recycle C&D waste from their site. Even in locales where commingled markets exist, the contractors may find the intangibles of separating waste to be a compelling reason to pursue that route. These intangibles include:

■ *Subcontractors do most of the work*: While it is true that separated recycling requires significant management efforts by the contractor, the physical labor of separating the waste is performed by subcontractors.
■ *Money saved is money earned*: In bid projects, contractors cut profit to the bone to get the job and work hard to regain profit through more efficient execution of the work. Money saved through source separation over commingling is potential profit for the contractor.

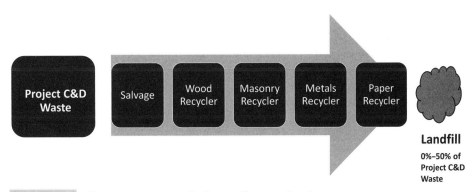

Figure 1.7   The source-separated recycling waste stream.

■ *Competitive markets*: Source separation usually offers more markets, and more opportunity to bargain for lower costs and better service. For high-value items such as metals, source separation lets contractors negotiate for greater income from waste of higher quality.

SELF-HAUL RECYCLABLES

Many contractors prefer to source-separate recyclables in piles or containers on-site and haul them to local recyclers or transfer sites themselves. A contractor may also be able to drop off commingled loads directly at a recycling company or end user. Depending on the labor costs in a particular area, this may be more advantageous than arranging with recyclers to pick up their particular waste at the jobsite. A second advantage of self-hauling is that the contractor is better able to maintain a clean and organized jobsite. Since he controls the schedule of pickups, he can assure that the need for overflow storage is limited or nonexistent, and that pickups occur when they have the least impact on demolition activity.

Before assuming he can self-haul, a contractor should examine closely each market's requirements for waste products. Some recycled product users—particularly transfer stations—may be reluctant to accept waste hauled directly by a contractor rather than a hauler they work with on a more frequent basis. Documentation may also be a concern. Municipalities, and even some private clients, may be reluctant to accept weight and haul tickets generated by a contractor for his own site. Because documentation from an independent weigher or hauler will usually be considered more reliable, such documentation may subject the contractor to less scrutiny.

The Recycling Method

Recycling construction and demolition waste, like any other construction activity, requires basic planning to be effective. In fact, the requirements of coordinating a successful construction waste recycling effort are quite similar to the types of planning required for basic construction project management, including scheduling, training, monitoring, and recordkeeping. Contractors who do not invest the time planning their recycling efforts will find it difficult to meet their goals for the project. This planning must include front-end training for employees and subcontractors, and a commitment from the contractor to enforce and document the progress of the plan throughout the course of construction. Planning should also include a commitment to recycling that provides the necessary equipment, containers, and management to ensure the program will be successful. (See Fig. 1.8.)

There are seven basic steps in creating and managing an effective jobsite recycling program:[11]

1 *Identify recyclable products*: The range of construction waste products that can be recycled is growing as new markets develop and grow. In larger municipalities, a diverse range of recycling companies already exists, enabling contractors to recycle as much as 95 percent of demolition waste. In rural areas, these markets may not

Figure 1.8   **Contaminated waste destined for a landfill.**
© 2010, Robert Asento, BigStockPhoto.com.

exist, or transportation distances and costs may preclude efficient on-site recycling. The first step for any contractor launching a recycling effort, therefore, is to identify which products on his jobsite are potentially recyclable based on the location of the project and the local market. Demolition contractors are often the professionals most able to help the contractor in assessing the recycling potential in the project area.

2 *Choose a recycling method*: Source-separated or commingled, individual programs run by subcontractors, a centrally-managed program by the general contractor, or some combination of these.

3 *Select recyclers*: Identify recycling markets, and the firms that service those markets, for each potential recyclable product on the jobsite. Dig into the details regarding pickups, costs, storage, and documentation. In particular, contractors should make sure they clarify the following points with a waste recycler:

■ What materials will be recycled?
■ How the weight or volume of recyclables will be documented?
■ Documentation of the requirement that all materials will go to a properly permitted facility.
■ Costs associated with the service, such as fees for container rental, hauling, recycling charges.
■ A schedule for picking up containers.
■ How waste containers will be labeled?

4 *Estimate the savings*: Details provided through contact with the recycling companies will provide the contractor with sufficient cost information to calculate the recycling savings versus disposal. Key to this step is an accurate estimate of the volume of demolition material, and a clear-eyed assessment of the value of that material. The savings in recycling comes from saving costs that would have otherwise gone toward landfill and tipping charges. Recycling does not make money so much as save the expenditure of costs that would otherwise have been incurred. This is an important distinction a contractor must make when he assesses the value of recycling a particular product. Even if some recycling hauler charges exist, they may well be less than the cost of container rental, hauling, and tipping for landfill waste.

5 *Train employees and subcontractors*: Contractors must train their employees to embrace and manage the program. Subcontractors who witness contractor indifference to the recycling program will be much less likely to participate diligently themselves. Include language in subcontractor agreements that they will recycle waste products from their work. Train subcontractors as well, including details of what they are responsible for recycling and how they are to accomplish it. Discuss the recycling program at every safety meeting or other regular meeting. Update all subcontractors and employees on the program's progress, and provide incentives or recognition to those who meet or exceed the goals.

6 *Monitor the program*: Check both the recycling area and trash containers regularly to make sure that all material is being separated properly, and recyclables are not being included in the trash being carted to the landfill. Insist on prompt and accurate documentation from both subcontractors and recycle haulers. Contractors can help to reinforce the recordkeeping by posting weekly updates on the recycling effort in the job trailer, and recognizing subcontractors who are on target with their efforts. If progress is lagging in the overall effort, offer incentives for subcontractors to look for other materials that can be recycled or to increase the level of recycling in previously identified products. The point person for a contractor's recycling program is the field superintendent, but she should receive ample support from the project manager and ownership of the company. To create a climate for an effective recycling program, the subcontractors and their employees must all realize the importance of recycling and the contractor's commitment to enforcing the spirit and letter of the program.

7 *Calculate final values*: Calculate the cost savings from recycling versus disposal using documentation from the hauler and subcontractors. This final step is where a diligent approach to documentation pays off for a contractor. Consistent recordkeeping (standard means of measuring volume or weight) and some basic research into the costs that would have otherwise been spent on landfilling will yield a reasonably accurate record of the savings achieved through the recycling program.

See Fig. 1.9 for a summary of the recycling system.

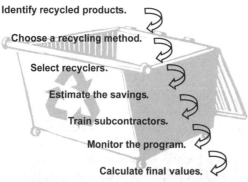

Identify recycled products.

Choose a recycling method.

Select recyclers.

Estimate the savings.

Train subcontractors.

Monitor the program.

Calculate final values.

Figure 1.9   The recycling method summary.

Buy Recycled Products

Markets for recycled products are growing daily and expanding from major urban and suburban areas to regional centers that service rural locales as well. Contractors may still find the markets limited in some areas for less common products they encounter in demolition projects. This will improve over time, but the marketplace for recycled construction and demolition waste can only expand as much as consumers are willing to buy products containing recycled material. Contractors hold the power to improve this situation—the power of the purse. Builders can improve the markets for their waste by purchasing recycled goods themselves. A wide range of building products is available with high recycled content. This information is readily available from manufacturers (many boldly advertise the recycled content of their products), enabling contractors to "buy green" with very little extra research or cost. Homebuilders, in particular, can market themselves as sustainable contractors by advertising the extent to which they use recycled products in the residences they build. General contractors, though often under product limitations imposed by the project specifications, can propose to the architect and owner that they allow substitutions for products with higher recycled content. The formula is simple: the more recycled products contractors and consumers purchase, the greater the market for recycled waste.

Summary

- Up to 95 percent of all construction and demolition waste can be recycled.
- Recycling saves money compared to landfill disposal.
- Source-separated recycling requires more management and labor than commingled recycling, but yields the greatest economic value.
- Contractors should organize and manage a jobsite recycling program by following seven basic steps: identify products, choose a recycling method, select recyclers, estimate the savings, train employees and subcontractors, monitor the program, and calculate the final savings.

References

1. Lennon, Mark. "Recycling Construction and Demolition Wastes—A Guide for Architects." Institution Recycling Network. August 15, 2009 <http://www.architects.org/emplibrary/Recycling_Guide_11-19-04.pdf>.
2. "Eco-Cycle's Ten Reasons to Recycle." *Eco-Cycle*. November 11, 2009 <http://www.ecocycle.org>.
3. Southworth, Matt. "Waste Generation and Commercial Projects: Median and Ranges." StopWaste.org. October 24, 2009 <http://www.stopwaste.org/home/index.asp>.

4. Turley, William. "How Much C&D Is Recycled?" Construction Materials Recycling Association. November 19, 2009 <http://www.epa.gov/epawaste/rcc/resources/ meetings/rcc-2007/day2/session-f/survey-cmra.pdf>.
5. Sandler, Ken. "Analyzing What's Recyclable in C&D Debris." United States Environmental Protection Agency (US EPA). December 3, 2009 <http://epa.gov/ climatechange/wycd/waste/downloads/Analyzing_C_D_Debris.pdf>.
6. "Memorandum of Understanding for Carpet Stewardship (MOU)." Carpet America Recovery Effort (CARE). December 5, 2009 <http://www.carpetrecovery.org/mou. php>.
7. Cassidy, Robert. "Getting Down and Dirty on C&D Waste Reycling." *Building Design & Construction Magazine*. December 27, 2009 <http://www.bdcnetwork.com/ article/CA6459382.html>.
8. Napier, Tom. "Construction Waste Management." National Institute of Building Sciences. November 27, 2009 <http://www.wbdg.org/resources/cwmgmt.php>.
9. Lennon, Mark. "Recycling Construction and Demolition Wastes—A Guide for Architects." Institution Recycling Network. August 15, 2009 <http://www.architects. org/emplibrary/Recycling_Guide_11-19-04.pdf>.
10. "About USGBC." U.S. Green Building Council (USGBC). November 29, 2009 <http://www.usgbc.org/DisplayPage.aspx?CMSPageID=124>.
11. "2008 King County/Seattle Construction Recycling Directory." King County (WA) Department of Natural Resources and Parks: Solid Waste Division. October 12, 2009 <http://your.kingcounty.gov/solidwaste/greenbuilding/documents/CDLguide. pdf>.

2

RECYCLING CONSTRUCTION & DEMOLITION WASTE: THE BASIC TOOLS

Basic Elements of Construction Waste Management

Managing construction waste on a jobsite is a crucial aspect of meeting the schedule and quality demands of the owner. Demolition, whether selective or whole, must be performed quickly and efficiently, readying the shell or site for the new construction work that will follow. For the contractor's field superintendent and project manager, the stakes are even higher. Layered on top of their other duties is a financial obligation to their employer to make the project profitable. Managing C&D waste effectively is an increasingly larger piece of the obligation these project principals must master, as both the requirements for recycling rates and the financial advantages of recycling increase. Part of this cost equation is recognizing that many opportunities for waste reduction exist beyond the pure demolition and construction work. Managing the schedule, ordering and shipping of materials, sitework, and other aspects of construction often can reduce the generation of waste. These reductions can and should be counted toward satisfying the contractor's recycling rate requirements for the project.

Types of Recycling and Reuse: The Rs

Recycling is many things. On a construction site, it is not merely the sorting and shipping to recyclers of demolition waste or new material scrap. A full and effective recycling program on a construction site includes a range of waste-reduction efforts, beginning well before the contractor even sets foot in the worksite. A comprehensive waste-reduction program also looks to maximize economies *within* the recycling effort itself, avoiding the creation of more waste through contaminated loads and sloppy sorting. The standard acronym offered for recycling is *RRR—Reduce, Reuse, Recycle*. This is sound advice for C&D waste management programs, but there are more fundamentals a contractor should look at as well. A complete program should contain the following elements:

1 *Reduce*: Waste source reduction strategies for suppliers and subcontractors.
2 *Reuse*: Review of material reuse possibilities and proposals from salvage companies.
3 *Recycle*: Detailed single-stream and source-separated plans.
4 *Return*: Take-back policies for material, including new and/or unused material, as well as waste cut-offs.
5 *Reprocess*: Alternative strategies to reprocess and reuse products on-site.

Following are some general tips for early planning across these five areas of waste management.

REDUCE: MATERIAL-EFFICIENT PLANNING

Contractors do not usually have much say in the overall design of the building, even in those situations in which they are acting as a construction manager in the planning stages to provide cost inputs. Nevertheless, the contractor can serve a valuable role in assisting the architect in identifying small dimensional changes that can result in savings. For example, by paying more careful attention to exterior dimensions of multistory facilities with a large exterior surface area on a small footprint, a contractor can achieve considerable savings through reducing board waste.

REDUCE: SOURCE REDUCTION

Both the generation of waste from construction activities and the amount of waste reaching the jobsite can be reduced. For example:

- When clearing the site, leave trees when possible; limit the extent of clearing and grubbing since vegetative and soil waste is a heavy component of C&D waste.
- Up to 12 percent of a project's construction waste stream can consist solely of cardboard. Direct subcontractors and suppliers to reduce extraneous packing and packaging.
- Purchase materials in bulk whenever possible; avoid individual packaging for volume purchases.
- Order finish or panel products with lower waste factors; control material usage more closely to limit waste.
- Use returnable containers and packing materials.
- Reuse nonreturnable containers on the jobsite as much as possible.
- Give away nonreturnable containers to local nonprofits or other organizations where reuse can be assured.

■ Contact distributors directly and urge them to reduce packaging from warehouse to site.

■ Urge subcontractors to limit the amount of waste they create from their trades.

■ Modify the floor plans with the architect to eliminate excess cut-offs.

■ Plan framing to maximize use of standard lumber lengths and board product sizes.

■ Use in-line framing, and increase stud spacing whenever possible.

■ Use scrap in lieu of cutting full new materials. Tell subcontractors to collect and keep scrap at cutting areas.

■ Collect paints and sealants from almost-empty containers; avoid disposing of useable materials simply because there is not enough in one container to finish the job at hand.

■ For materials that are heated, mixed, or subject to spoilage, limit preparation of these materials to quantities that can be installed within their expiration times. Work in smaller batches to reduce the necessity of disposing of expired or spoiled materials.

■ Provide protection or appropriate storage for materials that degrade when exposed to heat, cold, or moisture to ensure they are not ruined prior to use.

REUSE

A great deal of miscellaneous framing, board, and insulation waste is generated on even small construction sites. Save these miscellaneous materials in a bin and urge subcontractors to use them for their closure and boxing needs in lieu of cutting new material. For example, workers can:

■ Reuse cut-offs for blocking, headers, bridging material, and wall blocking.

■ Reuse plywood for forms, air barrier boxes for electrical outlets, panels, and plumbing fixtures.

■ Use crushed masonry material as fill for slabs, foundations, or paving.

■ Salvage materials such as sinks, tubs, wood floors, from demolition projects.

RECYCLE: SINGLE-STREAM

Single-stream recycling boosts the collection rate but ultimately causes more material to reach the landfill through the recycling center. On the jobsite, use single-stream recycling to limit the affected site area through fewer and larger containers. Under this system:

■ Clearly note acceptable and nonacceptable materials for single-stream collection.

■ Keep one or two large containers on-site; schedule frequent pickups.

RECYCLE: SOURCE SEPARATION

Source-separation economies require careful site management of containers, both in size and number. Market factors determine how material must be sorted on the site, and often how large a container can be used to store them. Taking advantage of a particular recycling market may also require some on-site processing, such as crushing or shredding. The contractor will need to decide if using this market is worth the cost of the extra labor and equipment rental required to use it. If source separation is selected as the recycling program for the jobsite, contractors should:

- Manage the number of containers on the jobsite carefully; limit the numbers and types.
- Require the waste generator to be responsible for placing materials in the appropriate container or cleanly stacking and binding the material outside the container on pallets.
- Provide subcontractors with rolling containers or other means to load recycled waste close to their work area.

RETURN: TAKE-BACK POLICIES

At the beginning of the project, the contractor and his subcontractors have a great deal of power in negotiating for favorable terms with distributors and material suppliers. This is an ideal time to include take-backs as part of the negotiation. Trade and finish subcontractors (such as carpenters, electricians, carpeting subcontractors, etc.) include a waste factor in all their orders. Although these subcontractors should be encouraged to limit their waste assumptions, they should also be required by the contractor to negotiate take-back policies with their suppliers, and document any new materials returned to the distributor for credit. This material may or may not count toward the recycling rate, depending on the strictness of the criteria for each project.

Actual cut-off wastes from various products are sometimes accepted by manufacturers for recycling. For example, some manufacturers take back waste such as carpet, padding, drywall, and vinyl to recycle into new product. This is becoming more common, and may be a subject of negotiation between the general contractor and the subcontractor or supplier; if this is the case, confirm that unused new materials can be returned to the distributor or manufacturer for credit. And, where the specifications permit, contractors should work with manufacturers who accept waste, and are willing to certify they recycle what they accept. To count this material toward his recycling rate, the contractor must ensure that documentation is available from the manufacturer certifying what percentage of the material is being reused in manufacturing new product.

REPROCESS: ALTERNATIVE METHODS

Innovative use of recycled materials on a construction site can often be limited in commercial or public projects. The specifications in these types of projects may allow little flexibility for a contractor to use field-processed recyclables as soil additives or in lieu of other specified products. Residential projects, and those constructed for owners with a clear interest in achieving high recycling rates, can offer greater opportunities for the contractor to suggest creative ways to recycle products on-site that would otherwise end up in a landfill. Architects can also be helpful in persuading the owner of the benefits of alternative recycling methods. Some opportunities exist for creative on-site recycling. For example:

- Chipped gypsum wallboard can sometimes be used on-site as soil amendment on lots prior to seeding or sodding.
- Chipped wood waste can be used on-site as mulch or compost.
- Crushed concrete can be used as aggregate for paving bases and other aggregate uses.

The Jobsite Recycling Center

The jobsite recycling center is the heart of a construction project's recycling effort. How well or how poorly it is organized will, in large measure, determine the effectiveness and profitability of the contractor's on-site recycling operation. In laying out an on-site recycling center, the contractor must consider and act upon a number of factors that affect its efficiency and operation:

- *Off-site road access*: The contractor should provide safe and well-marked access from the main road.
- *Maneuvering clearances*: The contractor should provide sufficient area for truck turn-around and drive-through, including for the loading and unloading of containers.
- *On-site access*: The contractor should provide clear and convenient access from on-site work areas, particularly for rolling bins or pickup trucks, to enable subcontractors to easily transport their waste to the jobsite recycling center.
- *Container size*: The contractor should match container sizes to markets, materials, and phases of the project.
- *Overflow storage*: The contractor should provide for overflow storage on pallets to allow for late or slow pickups.
- *Security*: The contractor should protect high-value recyclables and salvage material from theft and protect against drive-by dumping of hazardous waste and other contamination of recycling containers.
- *Surface stabilization*: The contractor should maintain a stabilized road base for truck and on-site traffic to the recycling zone.
- *Soil erosion and sedimentation control*: The contractor should maintain soil erosion fencing around the recycling area, and ensure that storm water runoff does not flow off-site.
- *Signage*: The contractor should clearly mark all recycling containers and overflow areas to ensure proper separation of materials.

Figure 2.1 summarizes these key considerations.

All these factors are important because failing to address any of them could end up costing the contractor money. Unclear signage that results in contaminated loads costs money; unsecured areas that allow drive-by contamination by the public costs money; insufficient clearances for loading, or hauler vehicles stuck in mud cost time and money. If a contractor is working on a site with sufficient area to support a full recycling zone, drive-through access for recycling haulers, and overflow areas for storage of recycled goods,

Jobsite Recycling Center Keys

❑ Ensure easy access for pickup.
❑ Use signage to designate material areas.
❑ Provide a stabilized base.
❑ Provide erosion and runoff protection.
❑ Secure valuable waste.
❑ Separate areas for separate materials.
❑ One-way traffic if possible.
❑ Overflow storage for excess waste.
❑ Keep area clean and organized.

Figure 2.1 Recycling center keys.

he should consider himself fortunate. This is often not the case, particularly on urban or occupied sites. In those cases, efficiency is the order of the day and field superintendents will struggle to keep order in the limited area available to them in which to operate. The basics of creating an organized and efficient recycling center apply even more in these situations, however, because the possibility of confusion, delay, and contamination is more prevalent. Here are some tips on setting up an *efficient* on-site recycling center:

1 *Minimize the number of containers*: Containers occupy valuable site area, and having too many containers increases the likelihood of confusion among the subcontractors and contamination of the loads. Contractors should aim to have one container on-site for mixed debris, and a maximum of two other containers for the specific wastes generated during each phase of the job. The specific waste containers required will change as the project progresses.

2 *Match container sizes to the material*: Different types of waste occupy more volume than others. Metal or conduit waste, for example, will require a much smaller container than wood or concrete waste. (See Table 2.1 for standard container sizes.)

3 *Consider site location and access when choosing containers*: It may be tempting to order a larger container to minimize transportation costs and the frequency of access, but if the result is recycling zone clutter because of limited area, it may not be worth it. Additionally, field superintendents do not need to be spending time dealing with haulers' pickup difficulties and clearance issues. An appropriately sized recycling area, with sufficient truck clearances and room to maneuver, is ideal. When that much space is not available, the best alternative is to limit the container size and compensate with more frequent pickups.

4 *Place containers close to work locations*: An advantage of source separation is that it does not rely on one big central container for all wastes. Smaller containers can often be placed close to the work. Contractors should also use wheeled hoppers, pickup trucks, or other means to enable subcontractors to load waste directly adjacent to their work area. This convenience tends to increase compliance and productivity. With limited site area, spreading access to containers can help to minimize clutter in the recycling zone. In locating containers outside a central zone, however, contractors must not forget the need to provide security against off-site dumping and protection for high-value recyclables. (See Fig. 2.2 for additional tips on keeping recycling areas clean.)

5 *Use signage liberally*: Clear signage, designating the recycling bins for each type of product, reduces contamination and increases compliance. (See this book's Online Resources for signage templates.) Prepare signage ahead of time, and rotate signage as the demolition and construction work progresses and the waste flow changes. For fool-proof compliance, link signage to laminated card handouts posted in each work area or provided to each worker on the site.

6 *Be flexible*: The types of wastes produced on a construction project change as the project progresses from clearing/grubbing through demolition to new construction. The superintendent needs to have a flexible approach to adjusting the recycling center bins and overflow to accommodate these changing needs, and must enable subcontractors to find a container for any potential recycled waste generated on the site at any stage of the project.

TABLE 2.1 STANDARD WASTE CONTAINER SIZES OF UNITED STATES AND UNITED KINGDOM	
UNITED STATES	
CONTAINER	**W × H × l**
2-cy dumpster	72" × 45" × 33.5"
3-cy dumpster	72" × 50.5" × 41.5"
4-cy dumpster	72" × 57" × 51.5"
6-cy dumpster	72" × 70" × 60"
10-cy roll-off	7' × 4' × 16'
15-cy roll-off	8' × 3.3' × 18'
20-cy roll-off	8' × 4.5' × 20'
30-cy roll-off	8' × 6' × 20'
40-cy roll-off	8' × 6.3' × 22'
Storage container	8' × 6.7' × 22'

Notes:
Add 8" to dumpster widths for side pockets.
Filled weight of dumpster should not exceed 1,000 lb.
Filled weight of roll-offs should not exceed 18,000 lb.
Filled weight of storage container should not exceed 6,000 lb.

UNITED KINGDOM

2-yd Mini Skip

Capacity		Height		Length		Width	
m^3	yd^3	m	ft	m	ft	m	ft
1.5	2	0.76	2'6"	1.2	4'	0.91	3'

4-yd^3 Midi Skip

Capacity		Height		Length		Width	
m^3	yd^3	m	ft	m	ft	m	ft
3	4	0.97	3'2"	1.83	6'	1.29	4'3"

8-yd^3 Large Builder Skip

Capacity		Height		Length		Width	
m^3	yd^3	m	ft	m	ft	m	ft
6	8	1.22	4'	3.66	12'	1.68	5'6"

10-yd^3 Maxi Skip

Capacity		Height		Length		Width	
m^3	yd^3	m	ft	m	ft	m	ft
8.85	10	1.5	4'11"	3.74	12'3"	1.78	5'10"

12-yd^3 Maxi Skip

Capacity		Height		Length		Width	
m^3	yd^3	m	ft	m	ft	m	ft
9.2	12	1.68	5'6"	3.7	12'2"	1.78	5'10"

Tip Box

- Include cleanliness provisions in subcontractor agreements.
- Conduct daily inspections of the recycling area.
- Discourage looters with security, signage, and lighting.
- Schedule on-demand pickups to avoid excess storage.
- Create overflow areas adjacent to dumpsters.

Figure 2.2   Keeping recycling container areas clean.

Team Management

Recycling can be an intricate operation. Project managers and field superintendents are accustomed to coordinating the details of construction. They are experts at laying out the schedule and critical events necessary to assemble a building from the ground up. They may not be as skilled at the process of *deconstruction*, however. The act of taking a building apart in a way that maximizes the ability to separate its original components is part science and part art. Layer on top of this the need to match products, containers, haulers, and markets, and the jobsite business of recycling can become complicated in a hurry. Demolition contractors can be a tremendous aid in the planning and execution of a recycling plan. They are becoming increasingly sophisticated in their understanding of how to efficiently remove materials from a building, find markets for them, and process them on-site to enhance their value.

THE PEOPLE

The first step in creating an effective jobsite recycling operation is to define the roles of principal parties in the project from a recycling point of view. The three primary organizations are the owner, the design professional team, and the contractor. It is important to note that each participant brings different motivations and skills to the effort:

- *The owner* has hired an architect and his consultants to design a building that meets his needs. He may or may not be environmentally sensitive, but he is almost always financially sensitive and sensitive to the market. When appropriate for the project, or mandated by local or municipal requirements, he will want to obtain the financial benefits of recycling and the marketing and public relations benefits of LEED compliance, or other certifications that allow him to claim a sustainable accomplishment for this project. Beyond the construction documents prepared by his design professional, the owner may seek to insert contract language that obligates the contractor to achieve recycling rates that exceed local requirements, or those required for LEED certification. This, of course, is a subject for negotiation. In smaller businesses the owner represents his own interests during construction. Larger businesses or corporations may employ facility or construction managers, or hire independent owner representatives to represent their interests. The contractor should always ask the owner to designate the Initial Decision Maker (IDM), the individual specified in standard contracts created by the American Institute of Architects as the person who is responsible for day-to-day decision-making for the owner.
- *The design professionals*—architects, engineers, and others—are much more likely to be motivated by true environmentalism than the owner, but they also are agents of the owner in meeting his facility needs. The construction documents (drawings and specifications) are the tools the architect uses to convey the contract requirements to contractors during bidding or negotiation. Architects who are seeking LEED certification for their projects will normally be quite specific in stating the recycling rate the contractor must meet to enable them to reach their LEED goals. These requirements

will turn up in specific specification language (see this book's Online Resources for an example) or less directly in the supplementary conditions or drawing notes. In cases in which municipal requirements govern recycling rates, and LEED compliance is not a factor, the contractor may find little or no guidance in the construction documents. Architectural firms are normally represented during construction by a project architect or project manager. In smaller practices, the principal of the firm may serve in this capacity.

■ *The contractor* enters the project, either through bid or negotiation, with a defined scope of work (represented by the contract documents), and a fairly open route to accomplish that scope. The means and methods of construction are his responsibility, and the course he chooses will, in large measure, determine the speed, quality, and profitability of the project. With regard to recycling, the contractor may well face contract or regulatory requirements that he achieve a certain recycling rate, but the way in which he achieves that goal is his responsibility. This flexibility requires keen judgments at the front of the project to balance time and money needs. The contractor is normally represented by two individuals in a project. The project manager is in charge of the project overall, and is responsible for financial management, contracts, and handling subcontractor and supplier issues. The field superintendent is responsible for jobsite management and day-to-day construction activities and schedule compliance.

Beyond these three major players—owner, design professional(s), and contractor—are the secondary members of the recycling effort. Although the major characters control the action, subcontractors and suppliers are the ones who are largely responsible for achieving the results. In particular, contractors should choose their demolition subcontractors carefully. Demolition subcontractors have become more sophisticated in their operations and knowledge of local markets for recycled materials. A savvy demolition sub can be a contractor's best friend in assisting him in assessing the market conditions and determining the likely recycling rate for the project.

These other subcontractors can bring more focused skills to the project's overall recycling efforts:

■ *Architectural salvage*: These subcontractors are adept at identifying the high-value items in a building that can generate actual revenue for a contractor. Ideally, the project manager should call in several different architectural salvage companies to provide bids to him, since different companies value the salvage based on their expertise and market knowledge. Some salvage companies, for instance, may be quite adept at removing and marketing bulk floor or wall finishes, while others will avoid this work and focus on architectural woodwork and other items more easily salvaged.

■ *Plumbing subcontractors*: Because of the nature of their work, plumbing subcontractors often are very aware of recycling markets for porcelain fixtures and miscellaneous metals. They may even be willing to manage that portion of the recycling market on certain renovation projects where bulk demolition is not desirable.

- *Electrical subcontractors*: Though less involved than plumbing subcontractors with marketing C&D electrical waste, some electrical subcontractors may be able to offer the contractor guidance on scrap metal markets, particularly markets that pay for high-value items such as copper wiring.
- *Recycling consultants*: Recycling consultants are becoming more commonplace across the nation as the markets recycling waste expand and become diverse. Recycling consultants are specialized consulting firms that exist to assist contractors in setting up a jobsite recycling program. These firms assess the potential recycling rate, and identify and contact markets for C&D waste. For a contractor embarking on his first recycling program, hiring a consultant can be a less painful way to learn the details of the recycling business.

Understanding the needs and motivations of all of these team members, as well as the expertise they bring to the project, is vital to a successful outcome of any construction recycling effort.

LEADERSHIP

The contractor's project manager is normally responsible for assembling the team who will carry out the recycling effort on the project. In consultation with the team members, he is also responsible for planning the program, leading the training, and tracking progress. The field superintendent is responsible for managing the on-site recycling center, daily monitoring of subcontractor compliance, arranging for pick-ups of recycled material, and collection of compliance documentation generated in the field.

The Importance of Training

Educating subcontractors about the goals of the jobsite recycling program, their responsibilities, and how the program will operate is key to the program's success. See Fig. 2.3 for tips on recycling training.

It is easy to overlook the importance of training. Construction workers are genuine professionals, and busy project managers and field superintendents may be likely to

- Teach subcontractors the materials to be recycled.
- Review procedures for storage and pickup.
- Convey record-keeping and reporting requirements.
- Review safety and security procedures.

Figure 2.3   Recycling training key points.

assume that they will understand and follow the mandates of a recycling program without coaching. This may be true in some, even most cases, but even an effort that is achieving its numerical goals may be ignoring one of the most important aspects of a recycling program.

Recycling is one of the great *feel-good* aspects of construction. Most workers take pleasure in seeing a building either transformed through renovation or rising out of the ground through new construction. They like being part of a team effort larger than their individual trade, and take pride in their portion of an overall accomplishment. Unfortunately, that reward—the completed work—always comes at the end of the project, and few tradesmen get to share in the glory or experience a sense of team achievement like that realized by the architect/owner/general contractor team.

A recycling program offers that opportunity. Occurring mostly at the beginning of a demolition project, recycling is a team effort that promotes an undeniable public good. When set up well by the general contractor, a jobsite recycling program can be a team-building, competitive, shared endeavor of the entire jobsite workforce. This is not to say that all will be perfect, or that the program will run without flaws. Some subcontractors will not participate fully, or at all, and the field superintendent and project manager will need to exercise the penalties they included in the general contractor/subcontractor agreement. Some workers, despite training and inducements, will never fully comprehend the problems they cause through contaminating a container with the wrong waste. Others will view the recycling program through a political or social lens they do not agree with, and use that viewpoint to denigrate it to their coworkers.

These things may happen, but they will happen with less frequency and intensity if the contractor launches a recycling training program with all the workers on-site, including his own personnel.[1]

ASPECTS OF A RECYCLING TRAINING AND MANAGEMENT PROGRAM

1 *Involve the subcontractors*: Ensure that the subcontractors and employees participate in the successful implementation of the plan. Educate them on the goals and details of the program. Require subcontractors to participate in the contractor's overall recycling effort, or to launch their own but provide suitable recordkeeping to the general contractor that documents their efforts. Include waste handling requirements in all project documents and agreements. Make it clear from the beginning that waste prevention and recycling is expected from all subcontractors and their suppliers.

2 *Promote and educate*: Teach subcontractors and employees how materials must be separated, where they will be stored on-site, and how often the materials will be collected and delivered to the appropriate recycling facilities.

3 *Document progress*: Let the subcontractors know how effective they have been by regularly posting the weights of material reused or recycled. Encourage suggestions from subcontractors and employees about identifying other materials that may be recycled, or how to create more efficient recycling procedures.

4 *Offer recognition and rewards:* Even hardened tradesmen appreciate recognition for a job well done, and offering rewards for exceptional performance by individuals or firms encourages greater participation and competition in reaching the goals.

5 *Keep the recycling area clean and organized:* Recycling efforts, and savings, may be wasted if recycling loads get mixed or contaminated with garbage. Haulers and recyclers will not accept contaminated materials, and may charge disposal fees to the contractor. Contamination of recycling is best avoided through information, education, and diligence. The contractor should conduct daily inspections of the recycling center, educating those subcontractors who are not sorting or storing properly, and punishing those who continue to do so. A more positive approach could include recognition of high-performing subcontractors and employees, including prizes as an incentive.

Transportation of Recycled Materials

Transportation costs can eat away at a contractor's recycling savings. In many markets, recyclers have no hauling ability of their own and expect the contractor to deliver the recycled products to their facility. They may or may not own containers. This places the contractor in a business that he does not want to be a part of: renting containers and operating a hauling service. While the contractor can certainly estimate the costs of both endeavors, he is in the business of building and would rather not have his field superintendent or project manager eating up valuable time in coordinating container and trucking services.

That is why contractors place high value on large recycling operations that provide their own containers and hauling. In a recycling endeavor, the most important goal to a contractor is preventing waste from heading to a landfill and sending it instead to a recycling facility. If he is under a contractual or municipal mandate to reach a certain recycling rate, the fact that contracting with a recycler saves little or nothing over the cost of landfilling the waste is less important than meeting the recycling requirement. See Fig. 2.4 for a hauler checklist.

The other criterion facing contractors in dealing with recycling transportation is the distance to market. In rural areas in particular, recycling processors may be too far from the jobsite, making transporting to this market uneconomical. The only options open to the contractor in these instances are to attempt to process the material on-site and to find a closer end-user who will accept the enhanced material. This is a dicey exercise, as the contractor's personnel must expend a great deal more time in researching unfamiliar markets that are normally serviced by the recycling company. The cost of renting the on-site processing equipment may also be as expensive as transporting to the distant market, so great care must be exercised in analyzing the best course.

Two types of common C&D waste haulers are:

- *Hauling trucks*: The workhorse of large recycling projects; hauls solid waste materials up to 44,000 lb, including solid waste, scrap, and bulky materials.
- *Trailers*: Trailers (transfer, roll-off, and walking floor) haul solid waste materials up to 80,000 lb, including various types of container storage.

□ **Cost**
    How much per load?
    How large of a load?
    Extra charges?
    Payment terms?

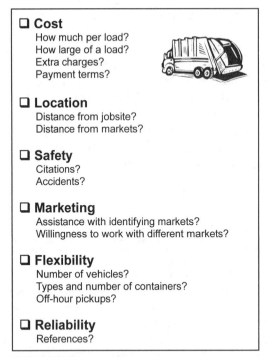

□ **Location**
    Distance from jobsite?
    Distance from markets?

□ **Safety**
    Citations?
    Accidents?

□ **Marketing**
    Assistance with identifying markets?
    Willingness to work with different markets?

□ **Flexibility**
    Number of vehicles?
    Types and number of containers?
    Off-hour pickups?

□ **Reliability**
    References?

Figure 2.4   **Checklist for recycling haulers.**

Hazardous Materials

The potential for discovering hazardous materials exists on any renovation and demolition site. Hazardous wastes are the bane of the recycling contractor, since there are few markets that will accept them, and ironically, relatively few landfills as well. The range of waste types ranges from fairly deadly (PCBs, mercury, and asbestos) to almost benign (universal and electronic waste). Hazardous waste regulations vary widely from state to state, though contractors facing environmental wastes on public projects may be forced to deal with federal regulations as well. Waste can be present in small thermostats, and wide expanses of floor and wall finishes. Some hazardous waste must be handled by certified industrial hygienists in a controlled environment, while other waste can be safely removed by trained demolition personnel using minimal protective equipment. Truly, the modern world of environmental hazards for the contractor is a witch's brew of confused rules, limited options, and varied threats.

Environmental consultants usually follow a hierarchy in the work of assessing the hazards present on a site or in an existing building slated for demolition:

1 A physical/visual site inspection
2 Process, purchasing, and inventory records

3 Material Safety Data Sheets (MSDS)

4 Past and present owners, supervisors, and employees

5 Town land records regarding past ownership and usage

6 Files maintained by the state or local authorities regarding activities and industries on the site

7 Analytical data (obtained either with field-monitoring equipment or through samples taken to a laboratory for analysis)

Among the hazards a contractor will encounter in demolition and renovation work, none is more prevalent than asbestos. Asbestos was widely used in many building products because of its tensile strength and chemical and thermal resistance. Asbestos, however, is extremely hazardous to workers and others who are exposed to it in any work environment where it is present in the air and may be inhaled. Asbestos is a known carcinogen that causes lung cancer and mesothelioma.

Because of this hazard, laws in most locales have been enacted to protect against asbestos hazards during demolition and renovation work. A key component of these laws is the requirement for building owners to inspect their buildings before construction, renovation, and demolition activities. Certain materials in buildings constructed prior to 1982 are presumed to contain asbestos until an inspection (and follow-up testing if necessary) proves otherwise. These materials include:

■ Surface materials (trowel or spray-applied surface treatments)

■ Thermal system insulation on pipes, tanks, and boilers

■ Flooring materials

A certified environmental consultant must conduct any assessment of presumed asbestos materials or any other suspected asbestos materials before they may be handled as non-asbestos materials. An assessment is not needed if a material is assumed to contain asbestos and handled as asbestos. Manufacturer or construction records may be used if the lack of asbestos content was documented when the material was installed. Previous surveys and abatement records may also be used if they cover the current work area.

Inspections must cover any material the contractor expects to be contacted or disturbed during work. Contractors must provide a written report of the environmental inspection findings to any contractor working in the facility, either in demolition or renovation work. He must also provide his own employees and any owners own forces working in the building access to the report and warn them of any materials that may be a hazard, as well as how to report suspected new hazards.

COMMON ASBESTOS MATERIALS

Special handling will be required for these building materials when asbestos is present:

■ Steam pipes, boilers, and ductwork insulation (thermal system insulation)

■ Resilient floor tiles (vinyl asbestos, asphalt, and rubber)

- Vinyl sheet flooring backing, and the adhesives used for installing floor tile
- Cement sheet, millboard, and paper used as insulation around furnaces and wood burning stoves
- Door gaskets used in furnaces, wood stoves, and coal stoves
- Soundproofing or decorative material sprayed on walls and ceilings
- Patching or joint compounds used for walls and ceilings
- Textured paints (sanding, scraping, or drilling these surfaces may release asbestos behind the paint)
- Asbestos cement roofing, shingles, and siding

See Chap. 4: Recycling Demolition Waste for a more extensive listing of asbestos and other hazards.

Not all areas of the country have statutes requiring an environmental survey prior to the start of demolition. Ideally, the owner has engaged an environmental consultant to assess and test for hazards regardless, and the design professional has clearly defined the responsibilities of the contractor in the contract documents. Unfortunately, this is not always the case. The owner may not have performed any testing to identify site hazards, and the architect may not have addressed the issue of hazardous materials in the construction documents in any way. Worse, the construction documents may contain veiled references or obscure notes making the contractor responsible for identifying and appropriately handling any hazards present on the site.

The contractor entering such a project therefore faces a dilemma. He cannot, legally or ethically, ignore potential hazards, but it is not prudent for a contractor to enter into an agreement where he accepts unlimited risk for dealing with undefined hazards. On any demolition or renovation project, the contractor must ensure that he is fully aware of the scope and nature of hazards he is responsible for handling under his contract, or that the owner recognizes that he is responsible for identifying hazardous materials and compensating the contractor for handling them. In a best-case scenario, the contractor is able to convince the owner to either arrange for environmental testing and consultation himself, or through his design professional. If absolutely necessary, the contractor can provide this service for additional compensation, though this is less desirable since it places full responsibility for identification, testing, remediation, or disposal entirely on the contractor. Whether by owner or contractor, however, professional environmental consulting should be retained to identify the hazards and recommend the proper handling.

Asbestos is such a common hazard in older buildings that contractors should have a plan on any demolition and renovation project for dealing with situations where it has either been remediated or is discovered or suspected during the course of the work. The following are basic recommended actions for a contractor in dealing with asbestos waste situations:

- Obtain a written asbestos report from the building owner or an environmental consultant hired by the owner.
- Provide the asbestos report to any subcontractors.

■ Look through the work area to check that asbestos has been identified and make sure the report is complete.
■ Make sure employees are aware of asbestos hazards in construction.
■ Make sure employees know about any asbestos materials on the jobsite that they may contact or disturb.
■ Report any suspect asbestos materials discovered to the building owner.
■ If asbestos materials must be disturbed or removed make sure a certified asbestos contractor handles the materials.
■ Prepare for accidental disturbances of asbestos materials—minor spills might be cleaned up with wet rags. Only use HEPA-filtered vacuums around asbestos materials.
■ Coordinate with other contractors and owners of businesses surrounding the project work area regarding asbestos issues.

Automated Equipment

Automated equipment can be rented by contractors to perform a wide variety of on-site operations that can convert C&D waste into a form that is either immediately usable on the construction site or more marketable and transportable to recyclers. As an example, chippers can convert wood waste, pallets, or tree/shrub waste into wood scrap or mulch. Wood scrap can easily be marketed to a variety of users, and mulch can be used on the jobsite or sold locally. Concrete crushers can convert concrete waste into aggregate for possible use on-site or sale to local markets. See Chap. 12: Resources for a description of all types of C&D waste-processing equipment. Following are some generic equipment types that are designed to either manage or process C&D waste.

ON-SITE PROCESSING EQUIPMENT

A range of on-site processing equipment is available to render large bulk waste items into crushed or shredded material, small pellets, or mulch that can be more affordably transported to more markets. Enhancing the value of C&D waste is one of the best ways available to contractors to boost their recycling rate. The difficult issue surrounding the use of such equipment on a particular project is whether it is worth the cost to rent the equipment and commit employee time to process the waste. In making this assessment, the contractor should keep two basic goals in mind:

1 *Can the material be sold to a local market without processing?*
2 *Can the material be affordably transported without on-site processing?*

If the answer to either question is no, then on-site processing may be valuable as a way to convert uneconomical waste into a form that can be a meaningful part of the jobsite recycling effort. When considering the use of on-site processing equipment, such equipment is most valuable if it can be operated by a single person and if it can handle a variety of waste types. Affordability of the equipment and the costs of its use, of course, cannot be ignored.

Here are some common types of value-enhancing on-site processing waste processing equipment, and their uses:

- *Compactors*: Crush waste with more than 65,000 lb (29,500 kg) of force and displace more than 175 yd^3 (134 m^3) per hour. Compactors handle drums, pallets, cranes, and bulky waste.
- *Pulverizers*: Punctured pieces of materials are dropped between rotating high teeth before being screened. Used to process gypsum board, industrial trash, and soft metals.
- *Separation systems*: Material is fed onto a vibrating screen in which the trommel sorts and discharges waste into different streams. These systems handle all types of C&D debris.
- *Horizontal balers*: Like separation systems, but designed with side-fed units. Can handle cardboard, metal, paper, and plastic.
- *Granulators*: Materials are broken up into pieces by rotors before being reduced into pellets by rolling teeth. Can handle plastics, rubber, foam, crates, and bins.
- *Grinders*: On-site grinding equipment is available for use on a number of products, including wood, plastic, and metal refuse. Truck-based grinders (such as the Grindzilla™) can handle a wide variety of product types, and are capable of separating metal from other nonmetallic waste.
- *Tub grinders*: Grind materials from 120 to 320 yd^3 (92 to 245 m^3) per hour from a top feeder with dual auger discharge. They handle typical C&D and land-clearing debris.
- *Trommels*: From a conveyor belt the fed material is screened and dispersed evenly, then outfed and stacked. Trommels handle yard waste and wood chips.

Summary

- Contractors should follow the basic Rs of sustainable waste management: reduce, reuse, and recycle.
- Aspects of a well-planned jobsite recycling center are: access and clearances, appropriate container sizes, security, soil erosion and sedimentation control, and clear signage.
- The recycling team should include the full range of project participants, including the owner, architect, contractor, demolition and other subcontractors, salvagers, and recycling companies.
- Recycling training and management programs should involve the subcontractors, promotion and education of all team members, recognition and prizes for achievement, and keeping the recycling area clean and organized.

- Asbestos-containing materials are the most common hazard found in demolition, and should be managed by a certified environmental consultant retained by either the owner or contractor.
- On-site processing equipment can reduce the bulk of demolition material, cutting container and hauling costs for contractors.

Reference

1. Gruzen Samton LLP with City Green Inc. "Construction and Demolition Waste Manual." New York City NY Department of Design and Construction. September 12, 2009 <http://www.nyc.gov/html/ddc/downloads/pdf/waste.pdf>.

3

RECYCLING NEW CONSTRUCTION
WASTE

New construction waste poses challenges and opportunities different from demolition waste. In working with new construction waste, the contractor has an opportunity to coordinate with suppliers and shippers to reduce the amount of waste entering the jobsite. This "early intervention" is equally effective for recycling efforts on the site, and depending on the project requirements, may well be counted toward the overall recycling rate of the project. At the very least, reducing the amount of waste coming onto the site reduces the contractor's costs and difficulties in managing the waste stream.

Reducing packaging and shipping waste requires flexibility and cooperation among the manufacturer/distributor, contractor, and subcontractor. Manufacturers package their products to reduce the likelihood of damage during shipping and handling. This protective envelope is normally expected to remain in place from manufacturer to distributor to end user. Assuming responsible handling and care, the manufacturer typically accepts responsibility for damage that occurs in spite of the protective packaging. In attempting to reduce the amount of packaging entering their jobsite, contractors are effectively asking manufacturers—or more likely, local distributors—to send them the product with less protection. This can be accomplished in different ways, such as using blankets or other reusable protection that will not enter the waste stream. The distributor will be concerned, however, as he is now being asked to guarantee a more exposed product. To further complicate matters, subcontractors are often responsible for picking up and transporting products under their contract to the jobsite. They must be as comfortable with the solution as the distributor. Following are some tips for contractors to use in accomplishing the goal of dealing with less product packaging and/or protection:

■ *Share the risk*: If the distributor and subcontractor are reluctant to remove packaging out of their fear of resulting damage, the contractor, as an incentive toward reducing the packaging, may offer to pay for any accidental damage.

■ *Provide reusable protection*: The contractor may purchase or compensate the sub-contractor for reusable protection that can be used to safely transport the product from distributor to jobsite.

See this book's Online Resources for a sample letter to vendors and suppliers requesting packaging reduction and other measures to reduce solid waste on the jobsite.

Waste Assessment

New construction waste assessment is, in some ways, more difficult than assessing the quantity of demolition waste. It can be very difficult to assess, sight unseen, the amount of packaging for products entering the jobsite. It can be equally difficult to determine the waste that will result from new construction operations. Skilled subcontractors usually estimate fairly closely the amount of material they will need for new construction, but the individual cut-offs from dimensioned lumber and board products can quickly accumulate to more than was anticipated.

Despite the millions of homes constructed in the United States, very little data has been collected on the amount of waste generated during the process of construction. In fact, only one comprehensive study is available. The U.S. Environmental Protection Agency (EPA) collected construction and demolition waste data for a 2003 study examining volumes of waste generated from new construction and renovation activities in the United States.[1] They found, not surprisingly, that waste amounts varied widely with the region of the country, size and type of the project, and type of construction. See Table 3.1 for a typical waste stream from residential demolition. For residential construction, the EPA collected data for National Association of Homebuilders single-family model homes constructed in four cities across the country. These similar homes, ranging from 2200 to 3000 (204 to 279 m^2) in area, generated an average of 4.44 lb of new construction waste per square foot (21.7 kg/m^2).[2] This figure tracked closely with the overall national average of 4.39 lb/ft^2 (21.4 kg/m^2) calculated by the EPA when they included other single-family- and multifamily homes in a range of sizes and locations.

Nonresidential new construction waste figures were estimated from a much smaller base of data, including only 11 projects ranging in size from a 5000-ft^2 (465 m^2) restaurant to a 1.6 million-ft^2 (149 million-m^2) office building. Although the EPA found a similarly wide range of results, and few patterns to rely on, the average waste generated by the full range of noncommercial projects they studied was 4.3 lb/ft^2 (21 kg/m^2), remarkably close to the residential waste figure. See Fig. 3.1 for new construction waste from commercial buildings.

In examining demolition waste, the EPA study concluded that waste generated during demolition typically runs 20 to 30 times more than waste created by construction activities. Their waste figures from the demolition of nine small-frame homes showed that an average residential demolition project generated 50 lb/ft^2 (255 kg/m^2) of waste,

TABLE 3.1   TYPICAL RESIDENTIAL WASTE COMPONENTS

Average solid waste generated by the demolition of a 2,000 SF (186 square meters) residence

MATERIAL	WEIGHT (lb)	(kg)	VOLUME (yd³)	(m³)
Drywall	2,000	907	6	4.6
Solid sawn wood	1,600	726	6	4.6
Engineered wood	1,400	635	5	3.8
Masonry	1,000	454	1	0.8
Cardboard	600	272	20	15.3
Metals	150	68	20	15.3
Vinyl (PVC)	150	68	1	0.8
Hazardous materials	50	23	—	—
Other	1,050	476	11	8.4
TOTAL	8,000 lb waste 3,629 kg		70 yd³ 53.5 m³	

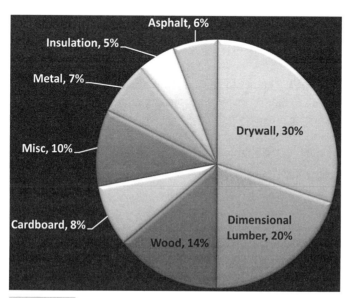

Figure 3.1   Typical new commercial construction waste stream.

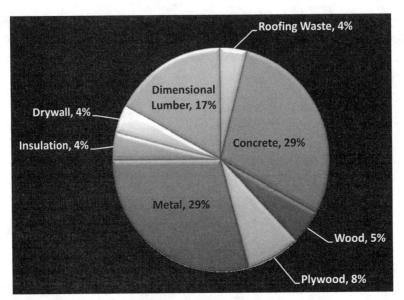

Figure 3.2 Typical demolition waste stream.

excluding concrete foundation waste. See Fig. 3.2 for waste stream information from a typical demolition project. A somewhat more extensive study of nonresidential demolition tracked figures for 11 structures, mostly located in the Western and Midwestern areas of the country. These structures ranged in size from 5700 to 2,200,000 ft^2 (530 to 204,386 m^2), and included a range of office, warehouse, institutional, retail, and military structures. The EPA study concluded that the average project generated 158 lb/ft^2 (772 kg/m^2) of waste. This figure matched exactly the results of a 2004 Association of General Contractors study.[3] See Chap. 4 for a discussion of demolition waste.

Residential remodeling and commercial renovation waste generation is difficult to assess on an average basis. The waste generated by a project such as replacing kitchen cabinets or a driveway can far exceed the waste generated by a simpler residential project like a bathroom or flooring project, yet the waste generated by the smaller project is much greater on a square-foot basis. In commercial construction, tenant retail renovations also vary wildly in extent and in the amounts of waste generated. Renovation includes a mix of demolition and new construction waste, both of which are totally dependent on the extent of work being performed. With those caveats, the EPA study noted the following waste generation values (based on a limited sampling of projects):

- Kitchen remodeling and room addition (500 ft^2, or 46.5 m^2): 19 lb/ft^2 of waste (4.9 kg/m^2)
- Bathroom remodeling (40 ft^2, or 3.7 m^2): 72 lb/ft^2 of waste (352 kg/m^2)
- Kitchen remodeling (150 ft^2, or 14 m^2): 64 lb/ft^2 of waste (313 kg/m^2)
- Whole house remodeling (1339 ft^2, or 124 m^2): 20 lb/ft^2 of waste (98 kg/m^2)
- New roof installation (1440 ft^2, or 134 m^2): 3.3 lb/ft^2 of waste (16 kg/m^2)

On the nonresidential side, the EPA surveyed renovation waste from nine office, retail, and hospital projects. Using the total values collected, the average renovation waste (demolition plus new construction) generated by all the projects they tracked was 11.4 lb/ft^2 (56 kg/m^2).

Waste Assessment Rules of Thumb

This rule of thumb assessment is based on average new construction waste and average percentages of waste types generated in the construction of an average wood-framed residence. Contractors should consider them a starting point for their own more detailed assessment of the waste stream (see Fig. 3.3). These values do not include masonry or concrete material foundation waste.

RESIDENTIAL NEW CONSTRUCTION CALCULATION

(*See this book's Online Resources for a spreadsheet calculator in English and SI.*)

Calculate the Total New Construction Waste (NCW) The gross building area $\times$ 4.5 = total new construction waste (NCW) in pounds

Calculate the New Construction Waste Weight Multiply the NCW by the following percentages to obtain individual material waste weight estimates:

- Dimensional lumber and wood products: NCW $\times$ 42 percent = pounds
- Drywall: NCW $\times$ 25 percent = pounds
- Masonry/tile: NCW $\times$ 11 percent = pounds
- Miscellaneous: NCW $\times$ 10 percent = pounds
- Roofing waste: NCW $\times$ 5 percent = pounds

Tip Box

- Estimate the Total New Construction Waste (NCW).
- Calculate the Weight of the New Construction Waste.
- Calculate the Material of New Construction Volume (NCV).
- Sort materials by market.
- Assess the container sizes needed and frequency of pickup.

Figure 3.3   Waste assessment procedure.

- Corrugated cardboard: NCW × 4 percent = pounds
- Metal: NCW × 2 percent = pounds
- Insulation: NCW × 1 percent = pounds

Calculate the Material New Construction Volume (NCV)   Convert the material weight to volume to estimate the number and type of containers required for the project.

- Dimensional lumber and wood products: material NCW/300 = NCV in cubic yards
- Drywall: material NCW/500 = NCV in cubic yards
- Masonry/tile: material NCW/1400 = NCV in cubic yards
- Miscellaneous: material NCW/350 = NCV in cubic yards
- Roofing waste: material NCW/1400 = NCV in cubic yards
- Corrugated cardboard: material NCW/100 = NCV in cubic yards
- Metal: material NCW/350 = NCV in cubic yards
- Insulation: material NCW/350 = NCV in cubic yards

Sort Materials for Local Recycling Markets (Varies by the Market)   Determine how to sort the materials on-site for the local markets, and create a schedule of anticipated container types and pickups. The containers needed for a given project will depend on the volume of materials, area available, and local situation. Contractors must also consider container security to avoid contamination through illegal dumping.

Sample Calculation for a 2000-ft² (186-m²) Home   Following is a sample new construction waste calculation for a 2000-ft^2 (186-m^2) home using the procedure outlined above.

$$2000 \times 4.5 = 9000 \text{ lb of new construction waste (NCW)}$$

Material Weight Estimates

- Dimensional lumber and wood products: 9000 × 42 percent = 3780 lb
- Drywall: 9000 × 25 percent = 2,250 lb
- Masonry/tile: 9000 × 11 percent = 990 lb
- Miscellaneous: 9000 × 10 percent = 900 lb
- Roofing waste: 9000 × 5 percent = 450 lb
- Corrugated cardboard: 9000 × 4 percent = 360 lb
- Metal: 9000 × 2 percent = 180 lb
- Insulation: 9000 × 1 percent = 90 lb

Material Volume Estimates (for Container Estimates)

- Dimensional lumber and wood products: 3780 lb/300 = 12.6 yd^3
- Drywall: 2250 lb/500 = 4.5 yd^3
- Masonry/tile: 990 lb/1400 = 0.7 yd^3

- Miscellaneous: 900 lb/350 = 2.6 yd^3
- Roofing waste: 450 lb/1400 = 0.3 yd^3
- Corrugated cardboard: 360 lb/100 = 3.6 yd^3
- Metal: 180 lb/350 = 0.5 yd^3
- Insulation: 90 lb/350 = 0.3 yd^3

Estimate Container Types and Usage (Assumes Source Separation)

- *Framing phase*: Wood waste: Order a 15-yd^3 (11.5-m^3) roll-off, hauled once; or a 6-yd^3 (4.6-m^3) container emptied three times.
- *Exterior and weatherization*: Order a 2-yd^3 Dumpster for roofing waste/order a 2-yd^3 Dumpster for masonry waste/store insulation waste in rolling container/store cardboard in rolling container.
- *Interior and finishes*: Retain a 2-yd^3 (1.5-m^3) Dumpster for corrugated cardboard waste (compact)/order a 6-yd^3 (4.6-m^3) Dumpster for drywall waste.

RESIDENTIAL NEW CONSTRUCTION CALCULATION (SI UNITS)

Calculate the Total New Construction Waste (NCW)   Gross building area × 2.0 = total new construction waste (NCW) in kilograms

Calculate the New Construction Waste Weight   Multiply the NCW by the following percentages to obtain individual material waste weight estimates:

- Dimensional lumber and wood products: NCW × 42 percent = kilograms.
- Drywall: NCW × 25 percent = kilograms
- Masonry/tile: NCW × 11 percent = kilograms
- Miscellaneous: NCW × 10 percent = kilograms
- Roofing waste: NCW × 5 percent = kilograms
- Corrugated cardboard: NCW × 4 percent = kilograms
- Metal: NCW × 2 percent = kilograms
- Insulation: NCW × 1 percent = kilograms

Calculate the Material New Construction Volume (NCV)   Convert the material weight to volume to estimate the number and type of containers required for the project.

- Dimensional lumber and wood products: material NCW/104 = NCV in cubic meters
- Drywall: material NCW/175 = NCV in cubic meters
- Masonry/tile: material NCW/485 = NCV in cubic meters
- Miscellaneous: material NCW/120 = NCV in cubic meters
- Roofing waste: material NCW/485 = NCV in cubic meters
- Corrugated cardboard: material NCW/35 = NCV in cubic meters
- Metal: material NCW/120 = NCV in cubic meters
- Insulation: material NCW/120 = NCV in cubic meters

Waste Types

CARDBOARD PACKAGING WASTE (OCC)

Cardboard and corrugated packaging (sometimes referred to as *old corrugated cardboard*, or OCC) comprises a fair amount of the waste generated on new construction projects. The term "cardboard" is used by paper and paper recycling businesses to mean the corrugated container board used mostly for packing and storage boxes. This term excludes paperboard, which is commonly used in cereal boxes, shoe boxes, and backing for writing pads of paper. Both cardboard and paperboard can be recycled, though paperboard has a more limited recycling market and may not be accepted with cardboard waste by all recyclers. With an increasing number of products shipped to jobsites in packaged form, the amount of cardboard waste can mushroom fairly easily. Fortunately, there are numerous markets in most areas for cardboard and paper waste.

Most cardboard markets restrict the levels of allowable contamination and will either not accept or pay substantially less for contaminated loads of cardboard. Contaminants are items that interfere with the recycling process. Cardboard contaminants include:

- Polystyrene foam
- Wood
- Plastic
- Metal
- Other nonsoluble materials such as plastic packaging tape, carton staples, adhesive labels, glue bindings, and kraft paper tape

Waxed cardboard is used for shipment of some food products, and cannot be recycled. It must be separated from nonwaxed cardboard. Additionally, old newspaper and office paper waste are considered contaminants to cardboard waste if they are present in large quantities. Wet cardboard can be recycled as long as it does not contain contaminants.

The construction of a single average facility does not normally generate enough cardboard waste on a consistent basis to interest a recycler in providing a container and hauling for the project, though there are exceptions. Contractors of large commercial facilities can generate substantial quantities of cardboard waste in certain phases of the construction. Particularly during the installation of interior accessories and furnishings, contractors may find that some markets are interested in working with the contractor to obtain the cardboard waste. Although the transaction may end up being a noncash transaction—the waste in exchange for providing of containers and hauling—the contractor should still investigate all the possibilities. Large sellers of cardboard waste recommend verifying the following specifics with potential recycling companies:

1 Location and capacity of each potential buyer
2 Current and historical prices paid per ton by the buyer for cardboard
3 Minimum and maximum quantities acceptable for pickup
4 Whether the buyer will pick up loose sheets or will require compacting or baling

5 Price differentials for different levels of service
6 Whether the buyer will furnish collection containers without charge
7 Contamination restrictions

Shipping alternatives for cardboard recycling normally include the following:

- Flattened and strapped to pallets: semitrailer
- Flattened and stacked loose: flatbed or box truck, compactor truck, or semitrailer
- Placed loose or compacted in Dumpster: compactor truck
- Placed loose or compacted in roll-off: roll-off truck
- Baled: flatbed, box truck, or semitrailer

POLYSTYRENE PACKAGING AND INSULATION WASTE

Expanded polystyrene (EPS) packaging and other forms of insulation waste pose a particular nuisance on jobsites. Polystyrene is commonly referred to as Styrofoam™, which is a trademark of the Dow Corporation. Polystyrene is routinely used by manufacturers in board form as a cushion to surround packaged products and protect them from damage. Although this product is an excellent material for recycling, it can be difficult to find local markets that will accept it. Though lightweight, it can be bulky and tedious to collect and store.

Currently, U.S. manufacturers recycle approximately 10 to 12 percent of EPS packaging each year. Although there are numerous end-use markets for EPS, the majority of EPS collected for recycling is used to make new foam products or remanufactured into rigid, durable products such as plastic, lumber, and molding trim. The Alliance of Foam Packaging Recyclers (AFPR) (www.epspackaging.org/info.html) maintains drop-off centers in 38 states. AFPR maintains a list of high-volume recyclers, who deal mostly with consistent sources of EPS or contractors who will be recycling a one-time large load. Large construction projects with a flood of EPS waste near the end may qualify for this service. Once a contractor has identified the closest collection site, he should call to find drop-off times and check to see what types of polystyrene material they accept. AFPR recommends the following before dropping off or mailing recycled EPS:

1 Ensure the EPS is clean and free of any tape, plastic film, labels, loose parts, or glued-on cardboard.
2 Empty coolers or other EPS cartons; food or medical waste is prohibited.
3 Check with the local EPS recycling center to see if they accept other recyclables from the jobsite.

AFPR also offers a mail-back program that allows consumers to send the polystyrene through the mail. The sender must pay the postage costs, but this may be less expensive than the tipping charges for disposal.

The Earth911.com (http://earth911.com) Web site provides lists of recycling centers for all types of products, but can be particularly useful in locating places to dispose of polystyrene packing waste.

One of the best methods of reducing EPS waste is to halt it at its source by contacting distributors and shippers and asking them to use reusable/returnable shipping containers and protection. If properly documented, these efforts can normally be counted toward satisfying the recycling rate requirements of the project.

Polystyrene Loose Fill (Peanuts) A related type of polystyrene waste is that used for loose fill packaging (often called "packing peanuts"). This material, when clean, can be offered to local mailing centers. It can also be recycled through a national organization. For the location of local recycling centers, contact the Plastic Loose Fill Council (PLFC) at www.loosefillpackaging.com or call the PLFC's Peanut hotline at 800.828.2214. The hotline is an automated 24-hour service that provides callers with the location of the nearest site that accepts loose fill packaging for reuse.

NONFERROUS METAL WASTE

In almost any market, a contractor can find buyers for nonferrous metal waste, one of the high-value commodities of the C&D waste world. Almost 40 percent of the world's requirements for copper are met through recycling, and far less energy is used in recycling nonferrous metals such as copper, aluminum, lead, and tin than is consumed in mining them from the ground. Widespread local markets exist for nonferrous metals; they are truly the most robust of the recycling markets. Contractors with enough waste (not usually the case on new construction projects) will be able to bid their waste to markets and negotiate the best combination of price and service.

On the Internet, Recycle Net (www.recycle.net) offers extensive listings of markets for nonferrous metals, including regularly updated scrap values for various types. Because the scrap metal market is so vigorous, a contractor who is inexperienced in dealing with scrap metals should consult with a demolition contractor or recycling marketer to avoid being shortchanged in dealing with this market.

Types of nonferrous metal waste scrap markets are:

- Scrap copper
- Scrap brass and bronze
- Scrap aluminum
- Scrap zinc
- Scrap magnesium
- Scrap tin
- Scrap lead
- Used/reusable nonferrous metals

Most of the nonferrous metal waste on a new construction site comes from electrical cable and conduit trimmings, plumbing copper pipe cut-offs, flashing cut-offs, and gutter/downspout trimmings. In many cases, the wasted material in these trades is very slight, and may be claimed by the subcontractor for his own recycling purposes. As always, the contractor should seek (or insist upon) the subcontractor's assistance in documenting his recycling efforts for inclusion in the verification of compliance with the project's overall waste management plan. Nonferrous metals markets will often offer containers and hauling as part of their services.

Because nonferrous metal waste is so valuable, the contractor should keep the containers locked after hours to avoid both theft and contamination of the load. On-site shredding or grinding of metal waste is an option for contractors with substantial volumes of waste who want to reduce its volume or render it more marketable. Given the value of nonferrous waste, this is rarely necessary for marketing, and may only be valuable in those areas where markets are so remote from the jobsite that transportation of bulk metals would be expensive.

FERROUS METAL WASTE

Ferrous metals are structural metals that contain iron. They are all magnetic, and are used in the construction industry for a variety of structural, piping, and connector purposes. Ferrous metal waste in new construction should be quite limited. The material is expensive to purchase and fabricate, and subcontractors are usually careful to order carefully to avoid waste. Steel structural sections are normally fabricated in accordance with shop drawings to exact dimensions. Cut-offs from steel angles, rebar, strapping, studs, hat channels, and other miscellaneous framing members vary by job, but are not usually extensive in either volume or weight. All miscellaneous ferrous metal can be collected in one container and easily marketed to scrap metal dealers, provided they are not contaminated with other types of debris.

Examples of types of ferrous metals are:

■ Cast iron
■ Cast steel
■ Construction steel
■ Free cutting steel
■ High-grade steel
■ High-strain steel
■ High-temperature steel
■ Low-temperature steel
■ Spring steel
■ Stainless steel

Second only to nonferrous metals, ferrous metals are high-value items for the contractor. Follow the same tips offered in the nonferrous section of this chapter in marketing and protecting this waste. HVAC equipment can be a significant source of ferrous metal waste on commercial projects. See Fig. 3.4 for HVAC equipment weights.

WOOD FRAMING AND BOARD PRODUCT WASTE

In wood-framed new construction, cut-offs from board, paneling, and dimensioned lumber products can add up quickly. Cornell University estimates the wood waste generated from new construction each year totals 6.7 million tons, of which they calculate

Warm Air Furnaces	300 lb (136 kg)
Electric Heat Pump	600 lb (272 kg)
Steam or Hot Water Systems	1,000 lb (454 kg)
Floor, Wall, or Pipeless Furnace	200 lb (91 kg)
Built-in Electric Units	200 lb (91 kg)
Room Heaters	200 lb (91 kg)
Stoves	200 lb (91 kg)
Fire Places	300 lb (136 kg)
Central Air	600 lb (272 kg)

Figure 3.4   Standard HVAC equipment weights.

88 percent can be recycled to a wide variety of markets. With the exception of treated lumber, this waste can be collected in a single container and marketed to scrap wood dealers to create mulch or for use as boiler fuel.

Cut-offs from finish lumber, including trim and flooring pieces, may be of interest to manufacturers who can reuse them in new finish products. This latter market is specialized and limited, however, and spending the time searching for buyers may not be worth the effort unless a considerable amount of this type of waste will be generated on a project.

Service companies rent out a variety of shredding machines for on-site use, which can perform a range of tasks to help manage wood waste. Shredders can reduce the volume of wood waste, remove metals, and grind the waste into product suitable for on-site use as mulching, or as more desirable boiler fuel. For projects involving a large amount of wood framing (and a correspondingly large amount of wood waste), the equipment and labor expense of reducing the waste to smaller quantities may pay for itself in reduced transportation and container costs.

GYPSUM WALLBOARD

Scrap drywall can be recycled back into the manufacture of new drywall. This is currently practiced by many wallboard manufacturers with their own postindustrial scrap. While a small amount of paper is acceptable in the mix, the majority of it must be removed. The percentage of recycled product used in new gypsum board typically is between 10 and 20 percent. Though not widespread, the recycling of new gypsum board scraps is available in a number of areas.

Another potential market for new gypsum board scraps is in the Portland cement market. Gypsum is an ingredient in the manufacture of Portland cement, where it is added to control the setting time of the concrete. In the manufacturing process, gypsum is added to the cement clinker from the kiln as a fine powder. The typical gypsum content of Portland cement ranges from 5 to 10 percent. Because the purity of gypsum in the wallboard is a major concern to cement manufacturers, all the paper and impurities (nails, corner beads) should be removed for use by this market.

Scrap gypsum drywall is often added to composting systems. While the paper face of the gypsum board will biodegrade as part of the compost, the gypsum itself will not break down in the mix, and will simply be incorporated into the final compost product. This gypsum board additive will result in a compost product that is rich in calcium and sulfur.

In selected areas, gypsum wallboard scraps may be ground and used as a soil supplement. This could be desirable in farm country, but also may be allowed on nonfarming sites as a soil enrichment tool. It is not appropriate to use demolition drywall; only use new drywall scraps that are free of nails, drywall compound, or paint.

Gypsum is a product used in agriculture as a fertilizer and as a soil amendment, since it does contain calcium and sulfur, which are both essential plant nutrients. When used in agricultural applications, calcium would normally be applied as a rate of 100 to 200 lb/ac (112 to 224 kg/ha). Sulfur would be applied at a rate of 20 to 50 lb/ac (22.4 to 56 kg/ha).

Crushed gypsum board, when applied as a soil supplement, would probably have higher rates than those normally applied for agricultural purposes. The need for calcium and sulfur nutrient in site soil depends on the soil type, the existing soil supply, and the contribution from other sources. Gypsum is not a liming material and will not increase the soil pH. In fact, large applications of gypsum may lower the soil pH slightly. The effect, however, is not long-term and would not affect crop growth.

Depending on the locale, the use of drywall in this manner may require soil testing and approval of the local soil conservation district or health department. WasteCap Wisconsin, Inc. has published a technical paper on this subject, with recommended application rates and other data (www.wastecapwi.org).

SITE RESIDUAL WASTE

The term *residual waste* has a changeable definition depending on the field under discussion. In industrial or mining activities, it is the waste deposited on the site after the conclusion of the work, whatever that may be. In logging operations, residual waste may be leftover limbs, sawdust, and tree branches. In industrial operations, residual waste may consist of anything ranging from toxic chemicals to barren soil damaged by stored materials. Assuming the contractor does not inherit any residuals on the site where he is to build, construction site residuals consist of all the site deposits that result from his construction activities. Traditionally, these consist mostly of clearing and grubbing material resulting from removing trees, vegetation, and ground cover present in the construction zone to leave it clear for final grading and landscaping in accordance with the contract requirements.

This once meant clearing virtually the entire site. Architects and civil engineers once considered it necessary to clear the entire site of native vegetation so it could be newly graded and landscaped as part of the scope of providing the client with a completely new facility—site and all. This thinking has changed, driven in large part by LEED

(Leadership in Energy and Environmental Design) and the desire to alter the natural landscape to a lesser extent during construction. This is for the good, as the reduced extent of site clearing and alterations of the natural landscape results in less site soil and vegetative debris that must be disposed of at the end of the project.

The prevalent trend is to keep this material on site, there being little market for mixed soil/vegetative mixes. The cost of transporting this heavy material can also be extreme, even when takers can be found. To retain site residuals on the site, the contractor should take several steps to make reusing this material easier later in the project:

- Limit the extent of site clearing. The extent may already be designated on the civil engineering or landscaping plans for the project, but if it is not, mark the boundaries of clearing with the site contractor.
- Separate vegetative matter from topsoil. This is not always affordable or possible on large sites, or those with difficult undergrowth. To the extent possible, however, require the site contractor to separate vegetative material as he works and stockpile it in a designated area for grinding.
- Stump grindings from felled trees create a large amount of wood waste. Avoid cutting down trees that do not directly affect the construction zone, even if they are shown to be removed in the construction documents.
- Consider renting an on-site sifting machine to remove stones, vegetative matter, and other residue from topsoil to allow it to be reused for final grading or landscaping.

Carefully plan and maintain site access roads and traffic paths. Police site traffic to reduce soil erosion and guttering from vehicle traffic, thereby reducing the amount of grading and soil required later to repair this damage. The demolition waste stream differs from the new construction waste stream in the far greater amounts of concrete and metal waste produced by demolition. For this reason, demolition offers much greater challenges to the contractor in managing the separation of jobsite waste. (See Fig. 3.5.)

 Compliance Connection

- Earn one-two points under LEED v.3 for Materials & Resources (MR) Credit 2.

- MR Credit 2 awards *one point* for diverting or salvaging *50 percent* of construction waste on a project.

- MR Credit 2 awards *two points* for diverting or salvaging *75 percent* of construction waste on a project.

Figure 3.5   LEED waste management certification credits.

Residential Construction Waste Management

Most of this book deals with aspects of waste management on larger commercial construction sites. Waste from residential construction, however, is a significant contributor to the national waste stream, and even small builders are facing municipal requirements to document their C&D waste recycling efforts. (See Fig. 3.6.)

Residential waste management presents the contractor with a number of challenges that are not present on larger projects. Single-family lots do not usually offer very much area for storing multiple recycling containers, requiring the contractor to plan carefully the number and size of containers he will retain on the site. Because the quantities of recycled materials are low, he will have a difficult time finding any recycler willing to loan him containers for the relatively meager amount of material to be gained. Neighborhood dumping (also known as drive-by contamination)—waste placed in a container by a party other than the builder or subcontractor—is a greater risk, as is the theft of any metal or high-value recyclables.

Residential waste management factors to consider[4]:

■ When measured by weight or volume, wood, drywall, and cardboard comprise between 60 and 80 percent of total jobsite waste.
■ Wood waste alone accounts for 40 to 50 percent of the residential construction waste stream. Of this total, as much as half of the wood waste can come from engineered wood products. Glues and metal connectors in these types of products

Figure 3.6   **Most residential construction waste consists of wood and gypsum board cut-off material.**

affect their suitability for recycling markets. Contractors should contact their local wood waste processors to determine what types of engineered products they accept.

■ Vinyl and metals are generated in small quantities, but have high recycling value.

■ Cardboard waste is an increasingly large source of solid waste on most residential jobsites as more products such as windows, appliances, and cabinets are packaged when they arrive at the jobsite.

■ Most construction wood waste is appropriate for recycling as long as it is unpainted and untreated. This includes dimensional lumber, plywood, oriented-strand board, and particle board without laminates. Lumber treated for exterior exposure is considered a hazardous waste in most areas.

■ Brick, block, and asphalt shingle waste may be insignificant in terms of volume, but can be important in terms of weight. The lesson: be careful of hauling and pickup costs for these materials.

■ For most home builders, the largest share of waste that could be considered hazardous is generated from painting, sealing, cleaning, staining, and caulking.

■ Drive-by contamination is a significant problem in residential construction. Drive-by waste can be as much as 30 percent of the total volume of waste generated from a residential site, and may expose the contractor to hazardous waste problems as a result. The lesson: lock Dumpsters or prevent public access.

Homebuilders and small contractors increasingly need to acquire basic expertise in residential construction waste management, as they deal with rising landfill costs and burgeoning municipal requirements to recycle construction waste. Contractors need to invest the time in performing basic research on recycling techniques and markets. In most cases, they will find their costs are reduced through their recycling efforts. Many cities and counties have developed recycling guides, aimed primarily at the homebuilding market. These guides often contain basic information on how and what to recycle, as well as local markets that accept recycled material. Subcontractors can often be a source of information as well because they are more attuned to what is happening in their particular industry and will be aware of local recycling opportunities that the contractor may not know about. Other sources of recycling market information include local builders' organizations and nonprofit housing agencies.

Following are some strategies for economically managing residential construction waste:

REUSE WASTE ON SITE

Where state and local authorities permit, contractors can grind wood and gypsum waste and incorporate it into the site soil prior to seeding or sodding. Wood waste may also be suitable as plant mulching material. Up to 65 percent of construction waste can be eliminated through this technique. The considerable savings in container, hauling, and landfill costs are partially offset by the need to rent a mobile grinding unit. Contractors must seek approval prior to using this method, as municipalities or other governing authorities may want to review whether water or soil quality will be affected by this practice.

CLEAN-UP SERVICES

Residential clean-up services in some areas double as recycling services. These services visit the construction site at planned intervals linked to the progress of construction. Container costs are drastically reduced as the pickups are frequent, and the normal process of separating construction waste into small containers is all that is required. The hauling service picks up the waste at the site and handles all recycling tasks, typically handling at least 50 percent of the total waste for new construction. They often charge by the square foot for their services, allowing homebuilders to estimate their waste disposal charges up front. These services exist in areas with a combination of high disposal costs and well-developed recycling markets.

MARKET WOOD WASTE

Residential wood waste can be used in composing operations, as mulch, agricultural animal bedding, landfill cover, in reconstituted building produces, and as an industrial fuel source for boilers. Contractors should develop a list of ready markets for cut-off waste from their construction sites, and shop for free hauling and containers. Some markets may reject waste from engineered wood products such as plywood, engineered structural members, or oriented strand board out of concern for the glues used in their production.

MINIMIZE CONTAINER USE

Residential sites often impose space limitations on contractors, limiting the size of waste containers. Even where ample space is available, contractors are well-served by ordering several small containers, or using small rolling containers and scheduling more frequent pickups. Large containers tend to encourage workers to fill them, are difficult to monitor, and lead to more contaminated waste and drive-by dumping by area residents. The standard 30-yd^3, roll-off containers can represent a big portion of a contractor's total disposal costs. For wood waste, a builder should look at avoiding containers altogether by storing the waste in a fenced or otherwise enclosed area and hiring a hauler who will pick it up.

Summary

- New construction waste assessment can be more difficult than demolition waste assessment, but is equally as important.
- The most prevalent types of waste resulting from new construction are: cardboard packing, polystyrene packaging and insulation, nonferrous and ferrous metals, wood, gypsum board, and site residual waste.
- Residential waste is similar, though the techniques for managing it require more flexibility and creativity.

References

1. Leigh, Nancy Green. "Construction & Demolition Debris Recycling for." United States Environmental Protection Agency (U.S. EPA). November 29, 2009 <http://cepm.louisville.edu/Pubs_WPapers/practiceguides/PG7.pdf>.
2. "Why Construction Waste Management?" Oikos.com. January 2, 2010 <http://oikos.com/library/waste/types.html>.
3. "Estimating 2003 Building Related Construction and Demolition Materials Amounts." United States Environmental Protection Agency. December 3, 2009 <http://www.epa.gov/epawaste/conserve/rrr/imr/cdm/pubs/cd-meas.pdf>.
4. Yost, Peter and Eric Lund. "Residential Construction Waste Management: A Builder's Field Guide." National Association of Home Builders. November 19, 2009 <http://www.docstoc.com/docs/376551/A-Builders-Field-Guide>.

4

RECYCLING DEMOLITION WASTE

The Site Audit

A waste management plan is the heart of any jobsite recycling operation. The plan cannot be prepared without a clear-eyed and conservative assessment of the actual field conditions affecting the demolition. This type of assessment is called the on-site audit.

An on-site audit is a critical part of a demolition recycling program. The audit enables all the major players in the recycling program to dig into the details of making the recycling program work. A team that includes the architect, contractor, demolition contractor, and recycler should convene on the site to assess what materials will be removed from the building, their estimated quantities, and whether their removal requires machinery or hand labor. An architectural salvage company may also be useful in helping to assess the potential for salvage of finish, cabinetry, or architectural woodwork items.

During the site audit, each member of the team should review the existing conditions in relation to the new construction plan, and solicit input from the other team members as to how their operations on-site will affect his work. The architect, for instance, may find that building features he thought could be retained and protected in place will actually need to be removed and reinstalled to replace the underlying substrate. The demolition subcontractor may discover that his work will be affected by architectural salvagers, who will precede him in the project and affect his schedule. The contractor may realize that finishes she originally felt held some salvage value do not, and the labor for their removal must be added to the demolition contractor's scope of work.

An individual with recycling experience can be invaluable during this phase. Whether that person is the architect, the demolition contractor, or a recycling facility operator invited for a consultation, she brings to the discussion the realities of the recycling industry. This industry, though it operates in parallel with the construction industry, often has a different outlook on the nature of the waste produced from a construction site. A recycler sees such waste as a commodity, a bulk product that he is paid to accept for processing (or in some cases she purchases it). The recycler does so with the understanding that the material meets his needs, which means it conforms to the requirements he

Tip Box

- Assemble recycling team:
  Demolition sub, architect, general contractor, recycling consultant

- Agree on scope of demolition.

- Review list of potential recyclables.

- Estimate rough quantities of recycled material.

- Assess salvage potential.

- Determine recycling zone location.

Figure 4.1   Demolition waste site audit tips.

has for what it can and cannot contain. Contaminated loads, or recycling containers that include materials that are unacceptable to the recycler, are trouble. They cause friction between the contractor and recycler and cost both organizations money in the form of wasted management and sorting time.

The primary purpose of the site assessment is to assess the possible. Regardless of the desired recyclables listed in the specifications, the site assessment must establish what *can* be done. Moreover, with all the principals in the field, it should establish *how* it can be done. The order of work, subcontractors involved, phasing, recycling zone setup, and other factors affecting the recycling effort should be addressed in the on-site assessment. Even if all difficulties cannot be resolved, at least the participants in the assessment are aware of the common challenges and should be prepared to help resolve them as the plan takes shape. See Fig. 4.1 for tips on performing the site assessment.

Following is a sample scope for an on-site assessment meeting:

- *Review of list of potential recyclables*: Which materials are not possible to recycle? Which materials have unknown markets?
- *Requirements*: What recycling rates, or other requirements, are mandated by the local municipality and/or the contract for construction?
- *Architectural salvage*: Are there items that a salvager may want? Are there installed items or finishes that must be protected and maintained, or removed and reinstalled?
- *Demolition*: How long will the demolition work? How many workers will be involved? What kind of equipment will be required?
- *Permits*: Has the demolition permit (or building permit) been granted? Have all environmental remediation permits been approved?
- *Insurance*: Review insurance requirements and documentation with the manager.
- *Safety*: What safety measures must be employed for structural stability and personal protection? Is personal protective equipment recommended for components of the demolition?

- *Temporary utilities*: Is temporary lighting or power required for the demolition or salvage work?
- *Hazardous wastes*: Has a survey been conducted to identify hazardous wastes on the property? If not, are there visible or suspected hidden conditions that require investigation and testing?
- *Recycling zone*: Where is the best site location for road and site access? How will materials be transported from the building to the recycling zone? What types and sizes of containers are needed for the demolition work? How should security and safety be provided?
- *Construction*: What types and sizes of containers are needed for the new construction work?
- *Training*: When should training be scheduled? What trades should be involved?
- *Documentation*: Who maintains documentation? How often should recycling plan progress be reported?
- *Recognition and rewards*: What type of worker/subcontractor recognition program can be set up to encourage participation? Who will manage this program?

More valuable materials, such as electrical wiring, nonferrous metals, and acoustical ceilings should also be reviewed, and their separate removal and storage should be planned for. Most importantly, the on-site audit also provides an opportunity for the design professionals and contractor to gain a common understanding of the opportunities and difficulties in meeting the goals of the waste management plan. The audit is the best time, occurring as it does at the very onset of the project, for the team to identify and resolve any conflicts between the paper goals and the realities of field demolition.

An additional review that must occur on selected projects is that of compliance with LEED (Leadership in Energy and Environmental Design) or other certification programs. For demolition waste management, three LEED credits come into play: Materials and Resources Credit 1.1 (Reusing building components), Materials and Resources Credit 2 (Waste management), and Materials and Resources Credit 3 (Materials reuse). The LEED program allows up to two credits for salvage, refurbishment, or otherwise reusing original building materials. See Fig. 4.2 for information related to the Building Reuse credit.

The review of possible materials that may be recycled is best organized around the Construction Specifications Institute (CSI) divisions of construction, which is the predominant format for most architectural specifications and construction estimating software. Following is a list of common C&D waste items suitable for recycling, listed by CSI divisions. This list is not exhaustive, but is included to provide a starting point for a team's assessment of their project's potential production of recycled products:

DIVISION 2: SITE CONSTRUCTION

- Asphalt
- Concrete paving, retaining walls, or slabs
- Paving units
- Retaining wall masonry

Compliance Connection

■ MR Credit 1.1: Building reuse—maintain existing walls, floors, and roof.

■ Up to three credits can be achieved through this requirement.

■ To earn the credits, the project must retain a minimum percentage of the existing building structure (including floor and roof decking) and superstructure (the exterior envelope and framing—but not the windows or roofing materials).

■ Building reuse points are awarded for the following percentages of existing building reuse:
  ■ 55% reused: 1 point
  ■ 75% reused: 2 points
  ■ 95% reused: 3 points

■ This credit is not applicable if the size of the project is being increased by more than 100%.

■ Remediated hazardous materials cannot be included in the calculation of reused materials.

Figure 4.2   LEED materials and resources credits 1.1: building reuse.

■ Wood from trees
■ Stockpiled soil
■ Topsoil
■ Stone
■ Site drainage structures and piping
■ Cobblestone

DIVISION 3: CONCRETE

■ Concrete footings
■ Interior floor slabs and poured concrete walls
■ Exterior cementitious panels

DIVISION 4: MASONRY

■ Brick for aggregate
■ Used brick for reuse
■ Concrete masonry units
■ Glass block

DIVISION 5: METALS

- Light-gauge- and cold-formed framing
- Structural steel members (angles, channels, and W-sections)
- Gutters and downspouts
- Metal siding or roofing
- Metal wall panels
- Flashing
- Metal decking
- Open-web joists
- Metal furnishings and cabinetry
- Metal tanks
- Metal handrails and grab bars

DIVISION 6: WOOD AND PLASTICS

- Dimensioned lumber framing
- Composite lumber products
- Heavy timber
- Glue-laminated beams and arches
- Pre-engineered joists and structural headers
- Plywood and composite board decking and sheathing products

DIVISION 7: THERMAL AND MOISTURE PROTECTION

- Expanded polystyrene insulation
- Loose-fill cellulose insulation
- Mineral wool insulation
- Mineral fiberboard insulation
- *Homasote*™ sound control boards
- Slate roofing
- Asphalt and composite roofing shingles
- Metal (steel or aluminum) roofing
- Asphalt-based roofing membranes

DIVISION 8: DOORS AND WINDOWS

- Wood and metal windows
- Vinyl windows
- Wood and metal doors
- Access hatches
- Basement entranceways
- Fixed interior lights
- Skylights
- Rolling doors
- Fire shutters

DIVISION 9: FINISHES

- Acoustical ceiling tiles
- Fiberboard products
- Resilient vinyl flooring
- Vinyl composition tiles (no vinyl-asbestos tiles)
- Ceramic and glass tiles
- Carpet and carpet backing
- Plastic laminate and solid-surface countertops
- Wood molding, flooring, and paneling
- Terrazzo
- Stone flooring
- Gypsum wallboard
- Rubber entry and service area mats
- Plastics of various types

DIVISION 10: SPECIALTIES

- Toilet partitions and urinal screens
- Toilet accessories

DIVISION 11: EQUIPMENT

- White goods: used appliances of all kinds
- Loading dock equipment

DIVISION 12: FURNISHINGS

- Furniture
- Fabric draperies and curtains
- PVC and metal blinds

DIVISION 13: SPECIAL CONSTRUCTION

- Tanks
- Interior catwalks
- Coolers and freezers

DIVISION 14: CONVEYING SYSTEMS

- Elevator cab and shaft metals
- Elevator and escalator motors
- Escalator components

DIVISION 15: MECHANICAL

- HVAC ductwork
- Equipment motors
- Tanks
- Plumbing piping
- Porcelain fixtures
- Sinks and tubs
- Hot water heaters and furnaces
- Boilers and radiators

DIVISION 16: ELECTRICAL

- Metal conduit and cabling
- Electric raceways
- Panel boards
- Transformers

See Fig. 4.3 for a checklist of questions to ask when considering recycling markets.

MIXED LOADS

Many areas have recycling processors who accept fully mixed loads of C&D waste at their sites. They use a combination of manual labor, front-end loaders or excavators with grapplers, and high-capacity mechanical processing systems to sort the waste by type before processing it further for the end users. Recycling facilities differ a great deal in how much hand labor or mechanized processes they use, so the efficiencies and costs of

- ☐ Do they provide hauling?
- ☐ How many materials do they handle?
- ☐ Where are they located?
- ☐ Are they properly permitted and licensed?
- ☐ What are their charges?
- ☐ What is their recycling rate?
- ☐ How quickly do they pay for metals or other positive revenue items?
- ☐ Do they have references?
- ☐ Are they insured to reasonable limits?
- ☐ Do they provide weight slips?
- ☐ Do they provide certificates of recycling or documentation of end uses?

Figure 4.3   Recycling market checklist.

such facilities also differ by a wide margin. Mixed debris processing efficiency can vary from 50 percent when manual labor is used to 95 percent efficiency when high-capacity equipment is used to perform the sorting. In the worst case, this means that fully half the load of debris being delivered to the mixed waste processor will end up in a landfill. This will quickly short-circuit any contractor's waste management plan, and he may be even more distressed when he sees that the rate varies up and down, depending on the sorting facilities' markets and the skill of their sorting personnel.

Furthermore, the recycling rate may not be accurately reported, or reported with sufficient detail to satisfy a municipal or project auditor. These facilities vary in their ability to provide verifiable documentation of their recycling rates for mixed loads processing. Contractors should take the time in assessing mixed load processors to verify both their efficiency and the quality of the documentation they can provide to him. The costs for mixed load recyclers are usually lower than landfill tipping fees, but manual recyclers may not provide much savings to a contractor seeking to lower his waste management costs through recycling. Mixed load processing fees always cost more than source-separated loads.

Materials commonly acceptable to recyclers in mixed loads (these should be verified with the mixed load recycler):

- Wood waste
- Tree trimmings
- Green waste
- Wood shingles
- Tile roofing
- Cardboard
- Ceramic tiles
- Metals
- Gypsum wallboard

Materials not acceptable at most mixed C&D waste facilities:

- Carpet
- Carpet padding
- Mixed composite materials
- Mattresses and furniture
- Acoustical ceiling tiles
- Plastic
- Pressure-treated lumber
- PVC pipe
- Food or organic waste
- Linoleum
- Paint
- Fabric
- Railroad ties
- Insulation/foam

The Market Audit

Pairing the C&D waste from a demolition project to the market is like arranging a business partnership. Recyclers have specific needs for and limitations on the types of waste they will accept. Contractors want to pay as little as possible in the case of bulk materials of little value, or gain as much income as possible from high-value products such as nonferrous metals or architectural salvage. Contractors have other needs as well. They need a reliable partner who will pick up containers when called or scheduled, and who will provide containers and hauling services reliably.

Although it is essential for a contractor to understand the waste products he will be trying to market, it is equally important for him to be aware of the limitations, preferences, and requirements of the local recycling markets. This awareness begins with an identification of potential markets. Contractors should:

- Research listings in the Kelly Blue Book, local yellow pages, or online sources (see this book's Resources appendix for region- and material-specific information and guidance on markets).
- Ask the demolition contractor for contacts and assistance on markets appropriate to the scope and type of project.
- Place an ad in a local newspaper or online, seeking market interest in the project. Be sure to list the expected types of waste the project will generate.
- Notify area plan rooms or construction bidding services about the project, including schedule and preliminary list of recycled products.

See Fig. 4.4 for typical uses for recycled products.

Beginning a business relationship with a recycler means exploring the boundaries. Both sides have minimum needs that must be met, and both have areas in which they have some room to be flexible. For a contractor, the main concerns are money and time. Other factors are important, but these factors override all others.

Demolition Recyclables

Demolition of an existing building offers society the opportunity to recapture the *embodied energy* of the structure—the resources originally used in the manufacture of the products and construction of the building. In areas with a healthy contingent of recycling markets, experts estimate that as much as 95 percent of the material generated in a building's demolition can be recycled. Even in areas with less-developed markets, conventional wisdom (and some statistical data) indicates that 50 percent of the materials can be recycled through a waste management program. Such a program requires diligence in closely assessing all the products in a structure that are available for recycling, and hard work in matching them to local markets.

Manufacturers are increasingly using recycled postconsumer waste in new products. The economic and marketing advantages of using recycled content are the main drivers

---

WOOD

Wood Fuel, Mulch, Bulking Agents for Composting, Manufactured Wood Products, Alternative Wood Fiber-Based Materials (e.g., Particle Board, Door Panels for Cars, Cements Additives)

CONCRETE

Roadbase, Fill Material, Aggregate for New Ready-Mix, Lime for a Neutralizing Agent, Rip-rap for Harbors (Large Pieces)

ASPHALT (Including Roofing)

Asphalt Patch for Roads (Cold-Mix) Pavement, Onsite Processing into Hot-Mix for Roads, Roadbase, or Fill Material

METALS

Reuse by Salvagers, Various Metal Feedstocks

GLASS

Reuse of Windows and Mirrors, Inert Granular Material Additive, Fiberglass, Reflective Beads, Glasphalt

DRYWALL

Soil Amendment (Gypsum), Cement Additive (Gypsum), New Drywall (Gypsum)

PAPER

Paper Fiber, Feedstock (Paper), and Animal Bedding (Paper)

RUBBLE

Aggregate for Fill for Roadbase, Construction Entrance Roads, Drainage Bed Material, Landfill Cover Material

---

Figure 4.4   **Typical uses for C&D waste.**

for this activity. Contractors can promote the greater use of recycled content in building construction products by themselves purchasing materials with recycled content. See Table 4.1 for an example of uses of recycled asphalt, brick, and concrete.

ARCHITECTURAL SALVAGE

Architectural salvage is the proverbial "low-hanging fruit" of recycling. Most contractors can readily identify existing building features that will have some value on the architectural salvage market—though there are surprises occasionally. The best course for a contractor is to contact one or more architectural salvage companies and ask for proposals. Various salvagers have unique specialties, and one may be aware of market opportunities that others are not. Occasionally, owners and design professionals will require the contractor to arrange for architectural salvage and credit the value of the contract to the owner. In other instances, the owner may contract with an architectural salvage company as an independent contractor under his employ (owner's own forces). Even in these instances, the contractor should seek to include the value of the architectural salvage in his own recycling plan. This is particularly the case when municipal or

TABLE 4.1 USES FOR RECYCLED ASPHALT, BRICK, AND CONCRETE

PRODUCTS MANUFACTURED FROM RECYCLED ASPHALT, BRICK, CONCRETE, ROOFING, AND OTHER MATERIALS							
Products Manufactured with Recycled C&D Waste Content	Construction and Demolition Waste Material						
	Asphalt	Brick & Concrete	Glass, Porcelain, & Ceramic	Roofing Shingles	Membrane Roofing	Catch-basin Grit	Processed contaminated soils
Bituminous Concrete Mixes	✓	✓		✓	✓		
Cold Patch Mix	✓			✓			
Crushed Gravel Substitutes		✓	✓	✓		✓	
Reclaimed Base and Fill						✓	✓
Hot De-Icer	✓			✓			
Manufactured Soil		✓					✓

specifications requirements, which require the contractor to recycle a given percentage of the waste generated on the project, are in force.

Salvagers should be allowed first into the demolition site once it is turned over by the contractor. Their proposal should indicate the visible materials they will be removing, and follow-up documentation from them should enumerate the types and quantities removed. Architectural salvagers should not be allowed to remove in-wall piping, electrical, or conduit. Bulk materials such as these are usually reserved to the demolition contractor for marketing under the overall waste management plan. As a general rule, architectural salvagers are only interested in appearance items that can be sold on the retail market to homeowners or commercial customers interested in vintage products (see Fig. 4.5). There are surprising exceptions, however, where salvagers will remove relatively common items such as water closets because they have identified a profitable market for their reuse.

Common products of interest to architectural salvagers include:

■ Cabinetry and countertops
■ Furnishings of architectural or historic merit
■ Architectural woodwork, including ornamental columns, standing and running trim, medallions, brackets, and fireplace surrounds, including any unusual or historic architectural features

Figure 4.5   Architectural salvage can include a wide range of products. *© 2010, Richard Thornton, BigStockPhoto.com*

- Ornamental plasterwork
- Antique plumbing fixtures and hardware
- Paneled doors and windows with historic glass or configurations
- Antique door hardware
- Ornamental radiators
- Glass block, etched glass, stained glass, or leaded glass
- Wood flooring and paneling, marble, stone, or ceramic tiles
- Ornamental or historic light fixtures

Architectural salvagers tend to be an independent lot, and were once frowned upon by the preservationist community. This has moderated somewhat, as more professional salvagers (or *deconstructers*, as they sometimes prefer to be called) are being seen as preservers of history. In dealing with architectural salvage and salvagers, contractors need to keep a few points in mind:

1 Some municipalities require that historic preservation organizations be given the opportunity to salvage items from a building of historic merit before demolition.
2 Preservation organizations may not have the tools and people to effectively salvage items in a professional and timely manner.
3 Professional salvage companies vary widely in expertise and capability. Check their references if they have performed salvage for other contractors.
4 Professional salvage companies will bid for salvage rights on some projects. If the value of the salvage is questionable, they may simply offer to remove it at no cost to the contractor.
5 Verify whether temporary power or lighting will be provided by the salvage company or the contractor during the architectural salvage phase of the project.
6 Stipulate the time period the salvager has to perform his work. Obtain up-front payment, and stipulate that salvage rights are void once the salvage period ends.
7 Ensure the salvage contract is only for visible and stipulated items. Do not permit salvage of hidden or discovered items.

FURNITURE AND FURNISHINGS

Furnishings run the full gamut of product types and offer a wide range of recycling opportunities, including some that may be unexpected by the contractor. Site furnishings can include such items as concrete waste containers and benches, or metal tree grates and bicycle racks. Concrete and metal products are readily recyclable with sorted materials from the building demolition. Plastic fabricated products may exist in relatively large quantities on some projects, including benches, picnic tables, deck furniture, and playground equipment. Depending on the composition of the plastic, these items can be recycled with plastic waste when they are separated from any metal or wood attachments.

Building furnishings of any value are normally removed by salvage companies. The types of furnishings salvagers would not claim are those that are damaged, contain suspected lead-based paint, or are integral to the building structure or framing and are therefore uneconomical or impractical to salvage. These furnishings are typically constructed or wood or metal, and once removed, can be added to the appropriate source-separated containers as long as they comply with the recyclers' requirements. Products with lead-based paint on them may fall under state guidelines for hazardous waste disposal, and must be handled accordingly.

ASPHALT AND BITUMEN

Asphalt can be recycled on construction sites in two principal ways. The first way is to simply use the existing paving as a base for the new paving. Design professionals may require testing, such as the California Bearing Ratio test (CBR), or other verification, that the existing asphalt surface is appropriate for use as a base for new paving.

Often, the elevation and drainage requirements will prevent use of the existing paving in whole as a new base.

In those cases, the existing asphalt can still be used, to some extent, by milling it. In a process called scarifying, a thickness of asphalt is removed and ground, resulting in a new base that can be overlaid with a new asphalt topping. In other cases where the existing pavement is severely deteriorated or cracked, paving contractors can mill the entire thickness for reuse as a new asphaltic base. Since the reconstituted asphaltic base may not measure up the requirements of a new installation, this option requires careful review with the design professionals to assess the suitability of the resulting paving for its intended use.

When on-site reuse of the existing asphalt pavement is clearly not appropriate, the material can be milled and sold to asphalt companies for use in the production of new asphalt. The material must be kept separate from other recyclables, and not mixed with soil or other site debris.

FERROUS METALS

Ferrous metals, such as structural steel, cold-formed, and light-gauge framing materials, can be marketed to scrap steel markets for use in new steel. There are typically few limitations to reuse of ferrous materials, though certain markets may request that structural steel be separated from lighter steel framing members (see Fig. 4.6). The contractor's dilemma in demolition is the need to remove gypsum board or other sheet products from the base of framing. Although much of the bulk removal can be accomplished with machinery, hand labor is required to clean up the light material to an extent acceptable to a scrap buyer.

Figure 4.6   **Metal waste has high value as a recycled product.**
© 2010, John Thielemann, BigStockPhoto.com

NONFERROUS METALS

Depending on the project, significant amounts of aluminum, copper, brass, and various alloys can be salvaged from electric, plumbing, and HVAC systems of buildings. Scrap recyclers readily accept recycled metals, and a contractor should be able to find competition for his waste in urban and developed areas. There are few limitations on recycling nonferrous metals, though the value is highest if they are separated by type. Even if the amount of waste does not warrant separation of the metals, scrap dealers will usually accept nonferrous and ferrous mixed metals.

AGGREGATE

Recycled aggregate is produced by crushing concrete, and sometimes asphalt, to reclaim the aggregate originally incorporated into it. Recycled aggregate can be used for many purposes. The primary market is road base, but it has a number of ancillary uses as well:

- In paved roads as aggregate base, aggregate subbase, and shoulders
- In gravel roads as surfacing
- As base for building foundations
- As fill for utility trenches

A roadway is built in several layers: pavement, base, and sometimes subbase. The pavement is the top, or surface layer, and is made of Portland cement or asphaltic cement. The base layer supports the pavement, and is made of aggregate base. A second base layer, the subbase, supports the base and is made of aggregate subbase. In road construction, the subbase layer contains more sand, and may contain more silt and clay than the base layer. As a result, this layer does not have as much strength as the base, but it is a more economical means of bringing the road up to the specified grade. Recycled aggregate can be used for the base and subbase in road construction.

Contractors may ship the concrete or asphalt to the processor in chunks, although this requires more hauling capacity and costs. For projects with a large amount of recycled material, the contractor may rent an on-site crushing plant. Portable crushing plants are available for transport to a jobsite. A crushing plant consists of a hopper to receive the material, a crushing jaw to break it into more manageable pieces, a cone or impact crusher to further reduce its size, a vibrating screen to sort the material to the required size, and a conveyor belt with a rotating magnet to remove metal contamination such as rebar.

Contractors must consult with particular recyclers prior to performing any on-site crushing of aggregate. Recyclers market aggregate to different markets, the most demanding of which are state transportation departments. Common terms used in marketing aggregate base are:

- Crushed aggregate base (may not include recycled aggregate)
- Crushed miscellaneous base (may contain crushed aggregate base or other rock)
- Processed miscellaneous base (may include broken or crushed concrete, asphalt, railroad ballast, glass, crushed rock, rock dust, or natural materials)

STRUCTURAL STEEL

In new construction, the quantity of structural steel cutoffs and waste is usually quite low. This ferrous metal material is valuable, and is therefore prepared in pre-engineered or stock sizes before being shipped to the project site. What waste is generated can be included in the ferrous waste container, and is easily marketable for income to the contractor.

CONCRETE

Formed concrete can be sold to aggregate markets in the same manner as brick and masonry. Heavily reinforced concrete walls form a particular problem, since they are not acceptable to most aggregate companies as recycled material. They must be separated from brick and masonry on the jobsite, and can be sold as mixed debris to specialized markets that can deal with the rebar/concrete mix. Lead-based paint applied to concrete is a hazard that must be identified and may preclude using the recycled concrete in aggregate mixes.

MASONRY

Demolished masonry is typically sold to aggregate markets for mixture with concrete and brick waste in aggregate mixes. Cinder block and concrete block are equally marketable. Architects are increasingly savvy to retaining existing masonry walls in renovated structures where they can serve some usefulness as shear walls, fire separation walls, or backer walls for new construction. The LEED benefits of retaining and reusing existing construction have driven much of this change, though the pure economics of minimizing the work of skilled trades such as masons is also responsible as well. Depending on the market, masonry may be mixed with concrete and brick waste for container pickup, though it is essential for the contractor to confirm with the recycling market what type of waste they will accept for their particular aggregate mixes. Reinforced masonry must be kept separate from pure masonry and mortar debris. The reinforced sections can be sold as mixed debris.

As with formed concrete, any masonry surfaces with lead-based paint applied to them must be treated as hazardous waste and may not be acceptable to aggregate recyclers.

BRICK

Brick waste from demolition can be reused in two ways. Historic bricks, or those with unusual character or coloring, may be valuable when cleaned of masonry for reuse in the same project or resale for other uses. Bricks with historic value (and like appearance) in a particular area are prime candidates for this type of reuse. The dilemma, of course, is the necessity to either clean the brick of its attached mortar (for the highest resale value), or find a buyer who will accept the brick waste as is (for lower resale value). Markets for used brick of this type are usually nonprofit organizations with free or low-cost labor that can absorb the high cost of restoring brick to usefulness. It is usually not economically advantageous for a contractor to attempt this on his own unless the project specifications require reuse of the existing brick from a building.

More often, waste brick from demolition can be sold to aggregate markets for mixture with concrete and block masonry to create various aggregate products. There are few limitations on this use, though glazed brick may not be acceptable to some markets.

WIRING AND CONDUIT

Copper wiring has high resale value, and for this reason most electrical subcontractors are already ahead of the game in marketing demolished material on their own. Similar to plumbing subcontractors, electrical subcontractors may or may not consider their ultimate sale of this commodity in order to make their initial bid more competitive to the general contractor. The contractor should verify early in the process, even to the point of stipulating during the bid period, whether or not the electrical subcontractor is responsible for marketing the wiring and conduit waste on the project. Though the weight value of this material is not significant, it is a contribution to the overall recycling rate of the project and the contractor needs to include this value as part of his waste management plan.

WOOD

Wood waste is the largest component, by volume, of the waste stream generated by construction and demolition activities. The U.S. Forest Service estimates 23 million metric tons of wood waste is generated through building demolition operations each year.[1] Another 30 million yd^3 of log residue are created from urban tree-cutting and trimming operations. In 1999, approximately one billion board feet of lumber was removed from residential wood decks. Wood waste generated on a construction site includes remnants and cutoffs from sawn lumber and board products, pruned branches, stumps, shrubs, and whole trees from site clearing and grubbing operations.

The markets for wood waste include:

- Feedstock for engineered woods
- Landscape mulch
- Soil conditioner
- Animal bedding
- Compost additive
- Sewage sludge bulking medium
- Boiler fuel

All these uses have similar processing requirements. The wood waste must be separated from other wastes, cleaned of contaminants and fasteners, and processed to meet the market requirements through grinding or chipping. The final use of the wood waste often determines how clean and consistent the feedstock must be, though intermediate recyclers may accept material from a contractor with metal contaminants as long as they can remove the contaminants themselves.

Dimensioned lumber can almost always be marketed to scrap recyclers who use it for mulch or boiler fuel. Contractors should verify whether the users accept the material with nails, screws, or assorted metal attachments. The labor to remove such items can be considerable, so it may be advantageous to contract with a company willing to accept the lumber as is, even if the cost is somewhat higher. Contractors demolishing older buildings with lumber of "full" dimensions may find a market in certain areas with architectural salvage companies or building supply companies who will purchase the lumber for reuse as exposed framing.

Floor and roof trusses are marketable as well, subject to the same concern regarding metal truss connectors being accepted by the market. Because they do not contain a significant amount of framing material, however, trusses do not contribute significantly to the recycling rate.

Wood trim, and other ornamental woodwork not taken during architectural salvage, can be sold in one of two ways. Flooring manufacturers may purchase clean, unpainted wood for use in new flooring and molding products. Scrap wood dealers will accept damaged wood, treated wood, and even painted wood for use as boiler fuel. Any painted wood surface should be cleared for lead paint ahead of time. Lead-painted wood must be disposed of as hazardous waste in an approved manner in the jurisdiction of the project.

Board products, such as plywood, oriented strand board, and glulam beams can almost always be sold to scrap dealers as boiler fuel. Board products have few other markets; heavier structural members such as long glulams or similar proprietary products may yield some other markets, but they are relatively rare and difficult to locate under deadline.

Contractors in certain areas may be able to market lumber or wood finishes directly to manufacturers, particularly in the furniture, pallet, or wood composites fields. These markets are not easily identified by those outside the particular industries, and tend to be demanding in their raw material requirements.

PLUMBING PIPES AND FIXTURES

Porcelain fixtures, when not suitable for architectural salvage, can be sold to recyclers who grind them for use as aggregate or decorative chips. This normally requires the removal of hardware and plastic components. Plumbing pipes can be sold to scrap dealers as part of mixed metals. Plumbing subcontractors are increasingly aware of the value of scrap metal, and the contractor is well-advised to coordinate with them early in the process (perhaps as early as bidding) to determine whether the plumbing subcontractor or the general contractor will be responsible for recycling of plumbing metals. It may well be to the general contractor's advantage to allow the plumbing subcontractor to manage this aspect of the recycling, as long as he contributes to documentation of the overall recycling compliance plan.

ASPHALT ROOFING SHINGLES

Each year, nearly 11 million tons of waste asphalt roofing shingles are generated by roofing contractors. Asphalt roofing shingles are readily accepted by recyclers for use in asphalt and other paving materials. Asphalt roofing shingles are made of a felt mat saturated with asphalt, with small rock granules. Organic shingles contain 30 to 36 percent asphalt, while modern fiberglass shingles contain 19 to 22 percent asphaltic materials. Common uses for recycled asphalt shingle scraps include:

- Asphalt pavement
- Aggregate base and subbase
- Cold patch for potholes, sidewalks, utility cuts, driveways, ramps, bridges, and parking lots
- Pothole patch
- Road and ground cover
- New roofing
- Fuel oil

Contractors can reduce their transportation and landfill costs through renting on-site machinery to grind shingles, remove nails, and screen the material for whatever the market requires. Ground shingles can be marketed to asphalt producers for use in hot-mix asphalt. To prepare shingles for use in new products, the shingles must be ground to a specified size and contaminants must be removed.

Asphalt shingles are processed on-site using a grinder. Depending on the equipment used, the first grinding pass may result in pieces of two or three inches in size. Secondary grinding may be required to make smaller pieces that are more acceptable to the market. Aggregate base recyclers, for example, may require ¾-in pieces, and asphalt pavement may require ¼- to ½-in (6.35 to 12.7 mm) pieces.

Shingles may be accepted by some markets with small amounts of metal or wood, including roofing nails and staples, but the contractor should verify this first. Shingles must be free of hazardous waste such as asbestos or lead flashing. However, the incidence of asbestos-containing shingles in roof tear-offs today is extremely low. The total asbestos content of asphalt shingles manufactured in 1963 was only 0.02 percent. By 1977, the percentage had dropped to 0.00016 percent, making the likelihood of asbestos contamination from roofing shingles very rare.

COMMERCIAL ROOF MEMBRANES

Single-ply roof membranes constitute a large surface area on many commercial and retail projects. Membranes can be recycled into asphalt and other paving materials, and may be accepted by a recycler with miscellaneous metals such as flashings and drip edges included. As is the case with other products, recyclers will not accept membrane roofing with asbestos or other hazards present in the material.

HVAC DUCTWORK AND MOTORS

HVAC ductwork is acceptable in mixed ferrous metals and mixed recycling, though it is not a large contributor when compliance is measured through recycled weight. Other HVAC equipment, including motors, is not easily marketable.

GYPSUM WALLBOARD

Drywall is the principal interior wall material used in the United States. It is made of a sheet of gypsum covered on both sides with a paper facing and a paperboard backing. Gypsum is calcium sulfate dihydrate ($CaSO_4 \cdot 2H_2O$), a naturally occurring mineral that is mined in dried ancient sea beds. Some commonly heard names for drywall are actually brand names; Sheetrock® is a registered trademark of U.S. Gypsum Company.

Most drywall waste is generated from new construction (64 percent), followed by demolition (14 percent), manufacturing (12 percent), and renovation (10 percent). Due to its weight, gypsum board begs to be recycled since it can be a significant contributor to meeting recycling quotas on practically any project. Most of these markets apply to scraps or cutoffs from new gypsum board. Unfortunately, the market for recycling of gypsum board material removed from demolition is limited. Drywall waste from demolition sites may be recyclable for nonagricultural markets. The following contaminants should be considered:

1 Nails should be removed before processing.
2 Tape breaks down in compost, or can be screened out.
3 Joint compound is made primarily of limestone or gypsum, and is not considered a hazard. However, if the structure was built before the mid-1970s, asbestos may be present in the joint compound.
4 Paint usually covers demolition drywall. Structures built before 1978 may contain lead-based paint. Lead can be detected with an inexpensive lead paint test kit. Drywall with lead-based paint cannot be recycled and should be disposed of properly. Mercury may also be a concern.

Georgia Pacific and U.S. Gypsum offer recycling of scraps from new installations in markets they serve, but only if screws, nails, and corner bead are first removed. Gypsum from demolition activities may also be used as a site soil supplement in certain areas, though the contractor must verify with local authorities whether, and to what extent, this is permitted. Other uses for gypsum board include:

- *New drywall*: Many drywall manufacturing facilities use postmanufacturer and/or postconsumer scrap drywall to produce new drywall.
- *Portland cement*: Gypsum is an ingredient in Portland cement manufacture. It is added to the cement clinker before the ball mill.
- *Application to land*: Gypsum provides a source of sulfur and calcium to crops. Gypsum can also improve the drainage and texture of clayey soils.
- *Compost*: Scrap gypsum drywall, or the paper separated from drywall, is often added as an amendment to composting systems.
- *Miscellaneous*: Other uses that have been proposed include animal bedding, flea powder, and various uses in construction products.

The two major objectives of processing are separation of gypsum from the paper and reducing the size of the gypsum material. Several processing methods are used for preparing gypsum drywall on the construction site for recycling, but a major problem associated with drywall processing on-site is dust production. Some localities may require air permits to grind gypsum in the field. Field operations deal with this problem by incorporating a mist to minimize the creation and spreading of dust. Several vendors market versions of self-contained drywall processing equipment, known as tub grinders. Many tub grinders operate using some type of grinder followed by a screening system. A dust collection system is typically a part of this system.

Given the difficulties of finding uses for drywall scraps, some of the best alternatives for waste reduction include generating less waste through more efficient techniques and retaining the waste that is generated on-site. Here are some examples:

■ Constructing standard-sized walls and flat ceilings
■ Ordering custom-sized sheets for off-standard walls
■ Finding drywall substitutes that are reusable, such as modular demountable partitions for commercial applications
■ Placing drywall scraps in the interior wall cavities during new construction

CEILING TILES

Armstrong World Industries has established itself as a leader in recycling acoustical ceiling tiles. Begun in 1999, Armstrong's program has recycled approximately 70 million ft^2 (6.5 million m^2) of discarded ceiling tiles since its inception. With no other markets for ceiling tiles, this material—estimated to be 10,000 standard Dumpster loads—would have ended up in a landfill. Armstrong posts information for contractors interested in recycling ceiling tiles on its Web site at www.armstrong.com/environmental. The Web site also contains a recycling cost calculator to assist contractors in estimating the cost difference between landfilling and recycling tiles.

Armstrong will accept clean and dry material, though with some conditions. They directly accept truckload amounts only (minimum 30,000 ft^2, or 2787 m^2), and may insist on testing a sample first for acceptability as a recycled product. The company may accept less than a truckload of tiles through a network of consolidators. To be acceptable, ceiling tiles must be mold and paint-free, and not be vinyl, fabric, or foil-faced, or contain any gypsum ornamental wood or pulp surfaces. They must be stacked on pallets and shrink-wrapped or tightly bound. Ceiling tiles do not have to be manufactured by Armstrong for recycling. A full list of acceptance criteria is available through the Armstrong Web site.

Recycling ceiling tiles contributes to LEED Materials and Resources credits MR 2.

CARPET

Renovation and demolition can produce large quantities of carpeting. Each year, more than two million tons of used rugs and carpet are generated from renovation activities in the United States. By volume or weight, carpeting can be a significant contributor to recycled material volume—if a market can be found. The ability to recycle used carpet exists, and the market is growing but is still not nearly as robust as the markets for other demolition materials. Contractors may find limited opportunities to recycle small amounts of carpeting on small renovation projects, and comparatively greater opportunities on large projects for which replacement carpet is being purchased. Used carpets may also be used as a component to produce other products such as auto parts, carpet padding, plastic lumber, and parking stops.

Affordable housing organizations may accept used carpeting suitable for cleaning and reuse in their projects. Some Habitat for Humanity affiliates accept like-new construction materials for affordable housing, though they typically require a minimum carpet quantity of approximately 800 ft^2 (74 m^2). Tandus Corporation, headquartered in Dalton, Georgia, operates carpet mills nationwide and accepts old carpet for reprocessing into the backing of new carpet products.

Old carpet can be broken down into separate components, which are recycled into several different new products. Several other manufacturers accept carpet for recycling, but under certain conditions. For recycling, the old carpet must be dry and mold-free. Even when old carpeting is accepted for recycling, the costs to the contractor can be high. Contractors have the most negotiating room with carpet manufacturers who are in line to provide the new carpet for a project. Carpet padding is more marketable, and may be sold to foam recyclers in most areas.

Carpet recycling is an evolving area, with the industry working to improve recycling opportunities. For additional information, contact the Carpet America Recovery Effort (CARE), an industry-affiliated organization working to develop solutions to the problem of used carpet entering the waste stream. CARE's goal is to divert 40 percent of waste carpet by 2012. CARE provides technical and financial assistance to entrepreneurs working to make products from postconsumer carpet, but in some cases may also have market information for contractors seeking recyclers in their area.[1]

INSULATION PRODUCTS

Insulation products occur on a project in various forms, only some of which can be recycled easily. Acoustical insulation panels are lay-in products with fiberglass cores. Armstrong World Industries will recycle these panels into new products if they meet certain conditions. Visit Armstrong's Web site at www.armstrong.com/environmental for additional information. The technology to recycle fiberglass batt insulation into new board products has been developed, but not to an extent that has proven to be economically feasible, so fiberglass batts presently end up in landfills. Old cellulose insulation can be recycled into new insulation. In fact, cellulose insulation is one of the industry's success stories, with more than 80 percent of new cellulose coming from recycled newspapers. Annually, insulation manufacturers produce more than 700,000 tons (635,000 metric tonnes) of new cellulose insulation.

Recycling of polystyrene board products is available on a limited basis. Rastra, Inc. (www.rastra.com), with plants in Columbus, Ohio, and Albuquerque, New Mexico, accepts old polystyrene insulation for recycling into insulated concrete form products. Rastra claims that their final product has more than 85 percent recycled content.

PAINT

Quantities of excess paint may become a problem for a contractor if he inherits a stockpile in a renovated facility or excess results from the vagaries of rejected work on a project. Latex paint can be recycled, and in fact such recycling is mandated in California. It can be reprocessed as a new product, mixed with other pigments or additives, and sold as a recycled paint. This service is not available everywhere, but its availability is well worth investigating for a contractor who finds himself with an excess of latex paint. California paint recyclers include: Eco-Paint, Kelly-Moore, and Dunn-Edwards Corporation.

NONFERROUS METALS

As noted in Chap. 3, nonferrous metals are readily recyclable in practically any project location. Whereas the supply of nonferrous metals on a new construction project will be clean and limited, the supply on a demolition project may well be *dirty*, but ample. Dirty in the sense that the metal will come in a wide variety of forms, such as brass body plumbing fittings, stainless steel attachments, copper piping and connectors, or cast iron piping. Local markets will probably be happy to accept all nonferrous metals as mixed loads, since the value is high anyway and they would prefer to sort for their markets. There is also no real value to contractors in attempting to grind this material on-site, since metal recyclers would prefer to degrease, ball mill, sort, grade, or densify this material according to their purchaser's specifications.

Types of nonferrous metals include:

- All types of copper: insulated wire, copper tubing, housewire, phonewire, Cat V/coax
- Aluminum/copper radiators
- Copper transformers
- Electric motors
- All stainless steel: 300 Series, 400 Series, turnings, and solids.
- All nickel alloys and tool steels: nickel, Hastelloy C, Hastelloy X, Inconel, Monel, Ni-resist, titanium, carbide
- All types of brass: yellow brass, red brass, meters, brass plumbing, fixtures, faucets, and fittings
- All bronzes: rod brass, rod turnings
- Auto radiators
- Aluminum: aluminum extrusions, siding, cast, aluminum dross, turnings, litho, aluminum-insulated wire, bare EC wire, aluminum cans, aluminum radiators, wheels, clips, old sheet
- Transmissions, window frames
- Exotic metals: lead, gold, silver, magnesium, zinc, platinum
- Batteries: lead-acid batteries, auto batteries (with uncracked housings), steel-encased batteries, UPS units, dry-cell batteries

FERROUS METALS

Like their nonferrous cousins, ferrous metals on a demolition site are plentiful and marketable. Even more so than nonferrous metals, they tend to come with unwanted attachments that can bedevil a contractor who does not read the fine print in his recycler agreement. Steel studs, for instance, must have drywall and wood blocking removed. Metal decking must usually be separated from its concrete topping. The same is true of steel pan stairs with poured concrete inserts. Some recyclers will accept mixed concrete and ferrous metals, but the contractor will pay for this privilege. Since metals are one of the few high-value recyclables, the tendency is to maximize that value to boost the recycling savings (and profitability) of the contractor.

Types of scrap ferrous metals include:

- Iron
- Galvanized
- Solids: #1 heavy melt, #2 heavy melt, #2 bundles
- Cast
- Busheling
- Clips
- Punchings
- Sheet products
- Engines
- Transmissions (which cannot contain any oil)
- Rebar
- Plate and structural steel
- Demolition scrap steel of all kinds
- Steel furnishings
- Appliances

Hazardous Waste

In the case of C&D waste, few markets will accept waste that contains known hazards, and will insist on assurances from the demolition subcontractor or general contractor that the product they are delivering is "clean" of hazardous materials. Several LEED categories exclude hazardous waste from the calculation used to satisfy the credit requirements, particularly in the Materials and Resources (MR) section most applicable to recycling and waste management (see Fig. 4.7). Examples of common hazards present in renovation and demolition projects include:

- Lead-based paint on a wide range of wood, metal, and plaster surfaces
- Mold growth on a variety of surfaces
- Pigeon or other bird guano
- PCBs in transformers
- Lead, chemical, or petrochemical contaminants in site soil
- Bacterial growth in plumbing and duct systems

Compliance Connection

■ Many LEED credits do not allow hazardous waste volume to be included as part of the calculation.

■ MR Credit 1.1 (Building reuse) and MR Credit 2 (Construction waste management) exclude hazardous waste remediation from the amounts allowed to satisfy these requirements.

Figure 4.7   **LEED and hazardous waste.**

■ Dead animal carcasses
■ Radon hazards
■ Fuel storage tank contamination
■ Abandoned septic fields, tanks, or cesspools
■ Railroad ties (creosote or chemical preservatives)
■ Phone poles (creosote or chemical preservatives)
■ Pressure-treated wood

See Fig. 4.8 for a listing of additional C&D hazardous wastes typically found in many common projects.

Asbestos-containing materials (ACMs) are particularly prevalent in demolition, and can be found in a wide range of products installed in buildings through the late 1970s. Abcov Conversion Systems LLC estimates that ACMs were once used in 3000 to 5000 different building products. According to the U.S. Environmental Protection Agency (EPA), asbestos may be present in the following products in existing building renovation or demolition projects:

■ Nonfriable asbestos in floor tiles
■ Asbestos-cement corrugated sheets
■ Asbestos-cement flat sheets
■ Asbestos-cement pipe
■ Nonfriable asbestos in ceiling and wall panels
■ Asbestos insulation or wrapping around boilers and piping
■ Asbestos binder in plaster
■ Roof coatings
■ Flooring felt
■ Millboard
■ Rollboard

Asbestos is, by far, the most prevalent environmental hazard encountered in construction, with perhaps as much as 300 million lb (136 million kg) of ACM waste created annually. The traditional method of dealing with asbestos waste on a jobsite is to

- Acids and Caustics
- Aerosol Cans
- Antifreeze
- Asbestos
- Contaminated Soils
- Fluorescent Lighting (Tubes and Ballasts)
- Fuel/Gasoline Kerosene
- Glues
- Herbicides
- Motor Oil and Filters
- Mercury
- Nickel-Cadmium (Ni-cad) Batteries
- Paint (Lead, Oil, Latex)
- Painted Wood
- Paint Thinners
- PCBs
- Pesticides
- Poisons
- Pool Chemicals
- Propane Tanks
- Railroad Ties
- Rechargeable Batteries
- Solvents
- Thinners
- Treated Wood

Figure 4.8   **Hazardous materials checklist.**

"bag, tag, and bury" the waste in approved landfills. With landfill space becoming more restrictive, and increasing community opposition to the transportation and landfilling of hazardous wastes, contractors are under more pressure to find new ways to deal with ACM materials.

In renovation projects, encapsulation and enclosure can sometimes be used to retain material in the building but render the material harmless. This is more likely for non-friable products such as tiles or boards than for friable products, where asbestos fibers could escape and be picked up by the building's HVAC system. Abcov Conversion

Systems of New York has developed an EPA-approved, portable chemical/physical asbestos destruction system that can be used on-site to render ACMs harmless.

MATERIAL UNACCEPTABLE TO LANDFILLS

Any waste that has been in contact with lead-based paint such as:

- Plaster and plasterboard
- Metal poles
- Concrete
- Painting equipment
- Wallpaper
- Mechanical parts
- Containers (cans, buckets)
- Lumber (siding, cabinets, shingles, etc.)

Any waste that has been in contact with petroleum products such as:

- Storage tanks
- Containers, filters (oil, etc.)
- Pipes
- Mechanical systems/machine parts
- Soil
- Absorbent material (such as vermiculite)
- Concrete
- Paper towels and rags

Any waste that has been in contact with friable asbestos material such as:

- Pipe insulation
- Broken/chipped floor tiles
- Asbestos-cement products that have been crumbled/pulverized
- Material containing friable asbestos
- Roofing material that has been cut with a saw

Any waste that has been in contact with polychlorinated biphenyls (PCBs) such as:

- Transformers
- Capacitors
- Electrical components
- Fluorescent lighting ballasts
- Any waste that has come in contact with any liquid containing PCBs

Any waste that has been in contact with solvents (such as waste commonly found in industrial plants, chemical plants, laboratories, or construction sites), including:

- Caulking compounds
- Paint thinner
- Containers (packaging)
- Pipes
- Filters
- Vats
- Pumps
- Adhesives
- Mechanical and/or machine parts (such as valves)
- Cement
- Cabinets (and/or shelving)
- Flooring (including wood, carpet)
- Tar
- Soil
- Glazing compound
- Storage tanks
- Absorbent material (such as vermiculite)

Any waste that has been in contact with preservatives (pentachlorophenol and creosote), such as:

- Railroad ties
- Utility poles
- Soil
- Containers
- Any mechanical part used in manufacturing processes

Any waste that has been in contact with pesticides and/or herbicides such as:

- Containers (packaging)
- Vats
- Soil
- Concrete
- Mechanical/machine parts
- Wood used in any storage area containing pesticides and/or herbicides
- Any equipment used for application of pesticides and/or herbicides

Miscellaneous waste such as:

- Lamps (containing mercury)
- Unpolished fiberglass (*Bondo*™, for example)
- Liquid waste (paint, paint thinner, etc.) and containers (paint cans, etc.)
- Caulking tubes

See Fig. 4.9 for typical universal recycling codes. See Fig. 4.10 for examples of universal waste.

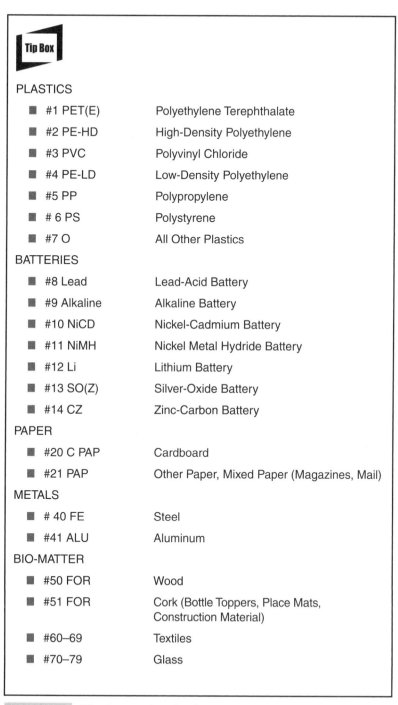

Figure 4.9   The international universal recycling codes.

- Ballasts
- Fluorescent Lamps
- HID: All Sizes
- Compact U-Shaped and Circular
- Batteries: All Types
- Mercury
- Relays, Switches, Ignitrons, and Thermostats
- Liquid Mercury
- Transformers
- PCB Capacitors
- Cell Phones

Figure 4.10   Universal waste types: E-waste.

Summary

- The site audit is a critical step in assessing demolition waste, and all key recycling team members should participate in the audit.
- The site audit should address: potential recyclables, recycling requirements, salvage potential, demolition, permits, insurance, safety, temporary utilities, hazardous waste, recycling zone location and configuration, construction schedule, training, documentation, and recognition/rewards.
- Demolition waste may contain a wide range of products with recycling potential. The contractor must take care to match the waste to the appropriate market, paying particular attention to the conditions imposed by recyclers on what constitutes acceptable waste.

Reference

1. "Frequently Asked Questions." Carpet America Recovery Effort (CARE). November 29, 2009 <http://www.carpetrecovery.org/faqs.php>.

5

REUSE OF EXISTING MATERIALS

In the recycling world, the term "reuse" often is used to refer to the salvage of building materials and their resale for use in another project. In this chapter, reuse means "keeping the original product on site." For example, reuse can mean reusing brick or steel exactly as it was originally intended to be used. It can also mean grinding wood framing into mulch or using the original cracked asphalt topping as a base for new paving. Reuse is the most beneficial form of recycling, and the one likely to save the contractor the most money, when compared to typical landfill disposal costs.

Building codes recognize the value, both economic and societal, of reusing existing building materials. The International Code Conference (ICC) includes specific language in the 2009 International Building Code (IBC) regarding the reuse of building materials. This section states: *2009 IBC Section 104.9.1: Used materials and equipment. The use of used materials which meet the requirements of this code for new materials is permitted. Used equipment and devices shall not be reused unless approved by the building official.*

The 2009 International Residential Code (IRC) contains similar language in Section R104.9.1. A complication to all the benefits of reuse is the basic economic agreement between an owner and a contractor. An owner who allows a contractor to reuse an existing product (and perhaps save disposal costs as well) will want a credit against the value of his contract. Contractors realize that reuse may in reality cost them more labor than landfilling or recycling, but these arguments do not always resonate with an owner who sees only that he has relieved the contractor of an obligation under the contract, and should receive something in return.

This chapter deals with the reuse of several common building materials, often found in significant quantities in many construction projects.

ABC: Asphalt, Brick, and Concrete Recycling

Asphalt, brick, and concrete are the literal building blocks of our society. They are, by weight, the most prevalent materials in the C&D waste stream, and the ones contractors pay the most to recycle. It stands to reason, therefore, that they also offer the greatest opportunity for savings to a contractor who can find creative—and inexpensive—ways to use them on-site (see Fig. 5.1).

This chapter is written around the concept that the best recycling is that which stays close to home. Reuse of demolition materials on-site is the most preferred means of recycling for several good reasons:

1 First, it negates the need to expend energy in loading, hauling, reprocessing, and transporting the material to an end user; all those transportation costs and effects disappear when a material is reused on-site.

2 Secondly, it fulfills the core concept of recycling: cradle-to-grave-to-cradle. Materials that are reused on their original site are restored to usefulness in the same facility in which they were originally installed.

3 Finally, reuse on-site is the most profitable way for a contractor to recycle. In the backwards finances of recycling, where markets cost money and *selling* means paying, reuse on-site usually costs less than any other option.

So what kinds of demolition material can be reused on the building site? A surprisingly wide variety. The catch is that reusing some of them requires the consent of the owner and the cooperation of the design professionals. If the owner and architect have stipulated a stiff recycling rate in the contract for construction, it indicates that

A *Asphalt*

B *Brick*

C *Concrete*

Predominant C&D Waste Materials
Most Commonly Recycled
Wide Range of Recycled Uses

Figure 5.1   Asphalt, brick, and concrete: the predominant recycled materials.

Compliance Connection

■ Earn one-two points under LEED v.3 for Materials & Resources (MR) Credit 3.

■ Points are awarded for salvaging, refurbishing, or otherwise reusing materials, based on the total value of materials used on the project.

■ Mechanical, electrical, plumbing components, and elevators are not included in the project value calculation.

■ Furnishings may only be included if they qualify for MR Credit 7: Certified Wood.

■ MR Credit 3 awards *one point* for *5 percent materials reuse.*

■ MR Credit 3 awards *two points* for diverting or salvaging *10 percent materials reuse.*

Figure 5.2   **LEED materials reuse certification credits.**

they are sincerely interested in the societal good that results from recycling, and that they should be cooperative partners in working with the contractor to reuse demolition materials. LEED (Leadership in Energy and Environmental Design) credit is also available for material reuse. See Fig. 5.2 for a summary.

ASPHALT

The American asphalt industry is one of the leaders in recycling. Approximately 73 million tons (66 million metric tonnes) of asphalt is recycled annually, constituting more than 80 percent of the amount of asphalt produced each year.[1] The Asphalt Paving Alliance estimated in 2007 that the asphalt recycled in the United States annually is more than twice the tonnage of recycled paper, glass, plastic, and aluminum *combined*. World demand for asphalt products is expected to increase by 2.6 percent per year through 2011, according to the Fredonia Group of Cleveland, Ohio, although growth in the relatively mature North American market will be below this level.

Paving companies reclaim the asphalt and aggregate, separating each for use in new recycled asphalt pavement (RAP). As much as 30 percent of new roadway paving may be comprised of recycled materials.

BRICK

Brick can be reused on existing sites in a couple of ways: as is or reconstituted into aggregate. Historic brick, or units with special character that matches other brick on the building or in the area, has enhanced value. It would be expensive, and in some cases impossible, to find matching or replacement brick. The problem is that existing brick has a troublesome attachment that comes with it—the mortar. Even in buildings with obviously deteriorated and cracking mortar joints, much of the mortar will remain attached to a demolished brick wall. Reusing the brick will require paying masons substantially more to remove the mortar as they work. A less expensive option is to pay laborers to perform this work, though the cost of doing so may prove excessive, and is only warranted to reclaim brick that has the rare value mentioned above.

For brick of lesser value, contractors can still retain it on-site by renting a grinder to render it as aggregate. The grinder will work on brick and mortar alike, and can produce aggregate appropriate for concrete slab or walkway bases, or for use around drainage pipes and foundation bases (if approved by the architect or engineer). When appropriate, brick aggregate may also be used alone, or in mixes with standard aggregates, as a base for site paving.

CONCRETE

Like brick, concrete can be ground on-site into aggregate that can be used for variety of purposes. Like brick, ground concrete waste can be used on-site as a base for new concrete and asphalt paving, drainage stone, or borrow pit fill. Unlike brick, concrete can be more problematic to reuse because it usually contains steel, in the form of rebar or welded wire mesh. Advanced on-site grinders can handle the grinding of concrete with embedded steel, and even sort out the metal after grinding. In selective cases, and with the approval of the architect and engineer, concrete slabs can be graded over and left buried on the site. The depth of fill and drainage are issues that must be reviewed before undertaking this measure. Most importantly to the contractor, though, is the fact that reusing concrete on-site can save a great deal of money in tipping fees to recyclers and in hauling costs for the weightiest component of demolition waste.

Concrete slabs may also be retained, without removal, as subbases for other work, including new paving, new concrete toppings, or finishes. This use is always subject to cooperation by the architect, and a mutual determination that the existing slab is suitable for use as a base for new construction.

National concrete repair services can clean, patch, and restore stained or cracked concrete to a like-new appearance in those situations where floor levels do not permit using a concrete topping.

Insulation

Clean, dry insulation products are often good candidates for reuse on projects. The dilemma facing the contractor in reusing insulation material is that it is light in weight; and recycling rates are normally calculated by the weight of materials, so reusing even a substantial amount of the insulation in a building will not dramatically improve the recycling rate when compared with the bulk of much heavier materials. Reusing these materials can, however, improve the bottom line, since they are bulky to store in containers and notoriously difficult products for which to find recycling markets.

Reusing insulation also requires the understanding and cooperation of the architect, who may be reluctant to reuse these materials for several good reasons:

1 Energy codes become stricter over time. The batt or board insulation that was code-compliant when originally installed is no longer, and the likelihood is that the architect has specified new and thicker insulation products with higher R-values. If he agrees to reuse the original insulation, it means the architect must either supplement it with other products (perhaps negating the savings of reuse) or reduce the overall energy efficiency of the building.
2 Batt insulation, in particular, is reduced in effectiveness through condensation or bulk water damage and compression. The architect may have valid concerns that, even if the product remained undamaged through years of use in a wall cavity, it will not remain so through the ordeal of being removed, stored on-site, and reinstalled.
3 Board insulation is normally glued or screwed to a substrate, or directly to wood or steel studs. Even if the material can be pulled off in complete sheets, screw holes or other damage to the panels may occur in the process of removal. This damage will compromise their effectiveness to some extent.

Board insulation that can be removed from framing without undue damage is a perfect candidate for reuse, providing the architect agrees that it meets the insulation requirements he has specified for the project. The labor required to remove and reinstall the insulation will be somewhat more than that required to demolish the existing insulation and install new panels, but not substantially more. Batt insulation that does not show signs of moisture damage is even easier to remove, and is another good candidate for reuse if the thickness and width work with the new cavity walls. Another opportunity for reuse of batt insulation is a form of modest *value-added* work for the owner. Even if the original batt insulation is not suitable for reuse in the exterior walls, the contractor can propose to install it on interior walls as low-level sound insulation. This extra work can usually be performed at modest cost, makes the owner happy that he is gaining an extra attribute in his facility, and saves the contractor the time and money of disposing of or recycling the insulation.

Structural Steel

Architects and structural engineers are usually well ahead of the contractor in identifying major structural steel components that can be kept and reused in a renovation project. They recognize the value of keeping these major structural members in place. On projects where they propose to remove steel W sections or other major members, or hidden steel supports are revealed during demolition, the contractor is perfectly entitled to ask the architect to instead retain these members in the project. There are at least three instances in which this may be advantageous to both the owner and contractor:

1 Where the architect proposes to close up a bearing wall that originally contained an opening supported by a steel member, there is usually little harm to leaving the steel section in place by encasing it in the newly closed wall.

2 Where additional load is being added to a steel section, it is often possible to supplement the load-bearing capacity of the member with additional web stiffeners, boxing the W section, or adding welded plates or channel steel to the bottom flange.

3 Where the opening is being widened, it may be possible to add the new steel below the existing member as long as the opening height requirements allow it.

Determining whether to reuse steel in place is a more complicated proposition for the contractor than the simple recycling equation. Scrap ferrous steel can be easily marketed (for income), so there is little question that steel removed from the project will contribute to the recycling rate. The more difficult pieces of the equation are the costs of keeping the steel in place, and convincing the owner that he is not entitled to a large credit as a result of this reuse.

Open-Web Joists and Miscellaneous Steel

Open-web (or bar) joists are designed and produced for specific spans, and are not readily adaptable to new spans, either shorter or longer. There is no effective way to increase or lessen their span. For this reason, architects may keep the existing joist framing in place and create a new open structure with structural steel W section and columns, enabling them to open up the framed space to an adjacent area with new framing.

Where the roof of an existing building is being raised as part of a renovation, it is ideal to remove and replace the existing joist framing at the higher elevation.

Miscellaneous steel is usually considered to be short span sections of light W sections, tube, or channel sections installed for particular or unusual uses. If the existing framing cannot be kept in place or easily supplemented, it is often not advantageous to attempt to reuse miscellaneous steel for other purposes. The cost and labor required to reengineer it, free it, rework it, and reinstall it almost always exceeds the cost of ordering new custom-designed steel members. However, miscellaneous steel can be recycled as mixed ferrous metals to scrap dealers.

Wood Framing and Heavy Timber

Wood framing is very adaptable for reuse in a project, provided the span and dimensions are compatible with the new use. Even lumber that has been notched or cut for piping can be repaired (in consultation with an engineer) for reuse in nonaesthetic applications. Framing lumber in older buildings may be of a full dimension, with more variable widths and thicknesses than sistered modern dimensioned lumber. As a result, it might not be appropriate for use in new applications requiring closer tolerances.

Heavy timber construction is often retained for its stability and aesthetic appeal. Fire-damaged heavy timber can sometimes retain enough strength to be acceptable in new applications. Even in those instances where the existing construction may not meet new code requirements (particularly lateral strength requirements for earthquake resistance), it may be possible to supplement the existing framing with additional members or brace it to bring it up to code.

The reuse of existing framing is not cost-effective or appropriate for all cases. Where substantial labor is required to remove, rework, and reinstall members, it may be least costly to remove the existing framing, recycle it as waste wood, and install new lumber that is appropriate for the use. A similar situation may exist where the structural capacity of the existing framing is unknown, or so severely below the requirements that leaving it in place and working around it would be much more difficult and costly than simply replacing it.

Finishes

Reusing existing finishes can be among the easiest, or most difficult, judgments in construction. This paradox occurs because it is so often not the finish itself that is in question, but the substrate or backing behind the finish. The call is easiest in floor finishes such as terrazzo, granite, marble, or ceramic tile installed on concrete slabs. Absent any objectionable staining or cracking, the flooring should be suitable for reuse if the contractor is confident the bond between the floor finish and slab is still reliable, and the grout between the flooring is sound.

Wall finishes are much more suspect. Even if the surface of the finish itself appears sound, the plaster or drywall behind the finish could be in the process of deterioration. Water or moisture damage from plumbing leaks, dew-point condensation, abandonment, or roof issues can result in moisture damage that occurs slowly but will inevitably result in deterioration of the finish substrate and require repair.

Finishes suitable for reuse:

- *Ceramic or stone over concrete walls*: Suitable for reuse, provided bond and grouting is sound.
- *Ceramic or stone over wood-framed walls*: Suspect for reuse, particularly if in a building that was left unheated or unmaintained over time.
- *Carpeting*: Rarely suitable for reuse.
- *Vinyl tile flooring*: Loses adhesion when left in unconditioned spaces. Rarely suitable for reuse, though other flooring products, notably carpet, may be installed over them if adhesion is sound.
- *Ceramic floor tile*: Suitable for removal and reuse (though mosaics are very difficult to remove and reuse).
- *Ceramic or stone flooring over concrete slabs or floors*: Suitable for reuse if bond and grout are sound (no hollow sounds when tapped, no excessive cracking at tile/grout edges).
- *Flooring over wood substrates*: Possibly suitable for reuse, depending on framing of floor and settlement of the substrate. Additional loading from new construction or occupancy can cause loss of bond of existing flooring.

- *Wood flooring*: Subject to bowing and warping if left in an unconditioned building and then re-climatized. May be removed, stabilized in a climate-controlled environment, and reinstalled.
- *Tin ceilings*: Generally suitable for removal and reuse. For use in place, verify attachments.
- *Acoustical ceiling tiles*: Not suitable for reuse.
- *Terrazzo*: Can be repaired, cleaned, and polished for reuse.
- *Marble*: Marble slabs used as wainscoting, toilet partitions, or for other purposes; can be patched, cleaned, and polished for a variety of new uses on a project.

Architectural Woodwork

Architectural woodwork is often a character-defining feature of a building, and is a prime candidate for reuse in a renovated structure. Where the architectural feature is so central to the character of the renovated building that it must remain, the contractor should warn the architectural salvager to leave it in place. In other cases, where there is some question as to whether architectural woodwork features should remain or be salvaged, the contractor should discuss the issue with the architect and owner. In heavy renovation, there may be little choice but to remove a desirable feature, store it for protection, and reinstall it in the finished facility. In the case of running trim, this inevitably leads to difficulties as the historic trim never installs as well or as tightly as new trim.

In other situations such as columns, medallions, brackets, or other stand-alone features, the return of an original piece of architectural woodwork to the new facility lends an air of stability and familiarity to the renovated building.

Glass

Recycled glass can be used in a surprising variety of new applications, ranging from aggregate based for paving to finish products such as terrazzo and countertops. For the contractor possessing a significant amount of glass product, identifying these specialty markets can be difficult. Demolition contractors with knowledge of area markets, or recycling consultants, can be very helpful in locating these markets.

Known recycled glass applications:

- *Container glass*: Reuse in new consumer containers.
- *Aggregates for bituminous materials*: Ground glass is acceptable in some states for use as an aggregate in asphalt paving.
- *Fiberglass insulation*: Fabrication of new fiberglass batts and board insulation products.

■ *Brick and tile manufacture*: Powdered glass can be used as a fluxing agent in the manufacture of brick and tile.

■ *Processed sand*: Recycled glass can be used in golf course sand bunkers, and as based for sports turf and artificial surface installations, for sports turf, dressing for artificial surfaces, and sand bunkers on golf courses.

■ *Concrete and cement*: Ground glass can act as a replacement for sand and pozzolan in the production of a wide range of concrete and cement products.

■ *Water filtration*: Recycled glass filter media can be used in water filtration systems in place of traditional sand filters.

■ *Grit blasting*: Glass grit is used as an inert abrasive in sandblasting operations.

■ *Glassphalt*: Asphalt containing glass cullet as an aggregate is called *glassphalt*, and has been widely used as way of recycling glass for over 50 years. Glassphalt is basically similar to normal hot-mix asphalt, except that as much as 40 percent of the aggregate is replaced by crushed glass. The financial benefits of substituting glass for conventional aggregate depend on the location and cost of local aggregates.

■ *Terrazzo finishes*: Glass can also be used to make a terrazzo-like product. Clear and colored glass is ground and tumbled, then melted and processed to produce a consistent color and size for use in terrazzo flooring.

■ *Countertops*: Vetrazzo, Inc. manufactures glass composite countertops that use approximately 85 percent recycled glass content. Vetrazzo states that their glass content comes from a variety of sources, including: curbside recycling programs, windows, dinnerware, stemware, windshields, stained glass, laboratory glass, reclaimed glass from building demolition, traffic lights, and other unusual sources.

Plastics and Composites

Recycled PET and HDPE is increasingly used in packaging by retailers and branded manufacturers for bottles and trays. Recycled plastic is also widely used in mainstream construction products such as damp proof membrane and drainage pipes and in ducting and flooring.

Walkways, jetties, pontoons, bridges, fences, and signs are increasingly being made from recycled plastic.

Street furniture, seating, bins, street signs, and planters are also frequently made from plastic. They are cost competitive and resistant to vandalism. Plastic film from sources such as pallet wrap, carrier bags, and agricultural film are made into new film products such as bin liners, carrier bags, and refuse sacks on a large scale. Polyester fleece clothing and polyester filling for duvets and coats are frequently made from recycled PET bottles.

Solid-surface products used for countertops, wainscots, window sills, and other uses may be kept in place, reused, or reworked. Since countertops are often installed over particle board bases, the quality of this substrate will deteriorate over time or when exposed to moisture. Plastic laminates, often referred to generically by the trade name *Formica*™, are not usually appropriate for reuse on a project. The material itself is difficult to remove from the substrate, and so inexpensive to purchase new, that it is not cost-effective to attempt to save it.

Other composite pieces that may be candidates for reuse are architectural elements such as columns, brackets, ornamental moldings, or running trim. When made from fiberglass or composite materials, these products are durable and perfect for reuse, as long as the architect agrees they are aesthetically appropriate for the project. If an architectural salvage company is involved, however, they will normally grab such items first.

Rubber (Tires)

Used automotive tires can be recycled for used in various applications, including:

- *Landscaping*: Shredded for landscaping mulch or reconstituted into new edging or other products.
- *Industrial*: Waste rubber is used in dock bumpers, door bumpers, and other products.
- Retreading of the existing tires (commercial or specialty tires, in particular).
- *Play industry*: Play surfacing, shredded and remanufactured into solid products.
- *Sports industry*: Sports surfacing.
- *Flooring*: Commercial and residential flooring products.
- *Equestrian*: Equestrian course obstacles.
- Fuel in cement kilns.

Asphalt Roofing Shingles

Asphalt roofing shingles can be ground on-site and used as an additive in the aggregate base used beneath site paving. The ground shingle material, in the right proportions, promotes compactability of the subbase. This use obviously is subject to review by the site engineer, and may not be permitted for heavy-use roads or parking, situations for which the specifications are stricter.

Several other potential markets exist for asphalt shingles. These include:

- Hot mix asphalt
- New roof shingles
- Cold patch
- Dust control on rural roads
- Temporary roads or driveways
- Fuel for boilers

Below is a detailed description of some of the more common uses for recycled asphalt shingles:

HOT-MIX ASPHALT

Hot-mix asphalt (HMA) is an established market for recycled shingles, and one in which their use is readily accepted because of the benefits recycling provides. These benefits include:

1 Reduced demand on virgin asphalt cement
2 Reduced demand for aggregate
3 Improved performance of hot-mix asphalt with recycled shingles

DUST CONTROL ON RURAL ROADS

Recycled asphalt shingles may be ground and mixed into the gravel used to cover rural unpaved roads, binding the stone, and reducing the buildup of dust over time.

ROOFING SHINGLES

The U.S. Department of Energy has estimated that a recycled content of up to 20 percent in new shingles does not affect the performance of the product.

COLD PATCH

When used as a cold patch, recycled shingles provide the advantages of longer life over traditional pure bitumen patches, likely because the fiberglass content provides better bonding.

TEMPORARY ROADS OR PARKING SURFACES

Recycled asphalt shingles may be used on construction sites for temporary roads, drives, or parking surfaces. The product is normally ground to 1/4 in (6.35 mm) and passed under a magnetic separator to remove all nails or staples before being ground and spread over a subbase.

AGGREGATE BASE

Recycled shingles blended with recycled asphalt pavement appears to improve the ability to compact the road subbase, making for longer-lasting installation.

FUEL

The burning of waste shingles as a boiler or process fuel has a long history in Europe, but far less use in the United States due to concerns regarding air pollution.

Gypsum Wallboard

See Chap. 4 for information regarding reuse of gypsum wallboard on-site as a soil additive.

Wood

See Chap. 4 for information regarding grinding wood framing for reuse as on-site mulch.

Reference

1. Oregon Bridge Delivery Partners, "Paving the Way." *Construction Demolition & Recycling*. November 14, 2009 <http://www.cdrecycler.com/articles/article. asp?Id=5120&SubCatID=73&CatID=4.>.

6

THE RECYCLING WASTE
MANAGEMENT PLAN

Create the Plan

The heart of any site recycling effort is the waste management plan. The plan outlines the basic collection procedures, monitoring, and marketing of the recycling operation. Most importantly, the plan establishes *profitability*, how much money the on-site recycling effort will save in comparison to sending the waste to a landfill. The plan is primarily a compliance document, generated to show the owner, architect, and municipality how the contractor intends to satisfy the relevant recycling requirements. It should be viewed more substantially by the contractor, however. The plan, in its purest form, is the skeletal outline of his goals in running a recycling program. It shows which materials are to be recycled, and approximately how much of each material the contractor expects to divert from landfills. It also shows the total waste expected to be generated from the project, and calculates the recycling rate. Like any plan, it will miss the mark in some areas and be relatively accurate in others. Because the contractor will track progress periodically as the project moves forward, the plan will enable the field superintendent and project manager to adjust to changing circumstances, missed estimates, and new opportunities. (Refer to this book's Resources section for examples of short- and long-form waste management plans.)

The recycling waste management plan has several purposes:

1 To demonstrate how regulatory or project requirements will be met
2 To identify the types and estimated quantities of waste that can be recycled
3 To determine how the waste will be controlled and stored on site
4 To determine companies to whom the waste will be marketed
5 To determine the project recycling rate

Define the Goals

The first step in crafting any plan is to determine the end goals. Increasingly, these goals are defined by the governing body (national and/or in the project's jurisdiction), or the project's specifications. Waste management plan goals are often stated in simple terms in zoning ordinances or approvals, or in the specifications generated by the architect or civil engineer for the project. Some examples are listed below:

- The waste management plan will recycle 50 percent of all construction and demolition waste.
- We will reuse or recycle 75 percent of all project waste.
- The contractor to limit landfill disposal to 30 percent of material waste generated by his activities.

Each of these statements defines the recycling goal for a project in terms of percentages of the overall waste. Defining the goals is simplest, of course, when they are provided to the contractor. A municipal ordinance or board approval referendum may provide the contractor with a clear statement of his end goal for recycling. Where the goal is clearly defined, the contractor can certainly exceed it, but he should do so for his own reasons. A requirement is exactly that—an obligation that must be met. A plan mandated by ordinance, resolution, or contract should be oriented toward setting an achievable goal that a reasonable contractor, operating in a normal business environment, should be able to achieve. Increasingly, counties and municipalities in urban and developed areas are mandating that 50 percent of C&D waste be recycled. In their view, at least, this goal is readily achievable. The market for recycled products would tend to support this viewpoint.

Define the Waste Products

The on-site assessment should have identified all the construction and demolition waste, by category, that will be produced by the project. With this list, the project manager can assess which of these waste types represent *products*, or waste that can be marketed to area recyclers. This step in the process is full of potential pitfalls, since a number of hazards may foil the project manager's initial assessment of how much of the project's waste he can recycle:

- Some recyclers may set stiff waste purity conditions that the contractor cannot economically comply with.
- Recyclers may charge more than landfill tipping fees.
- A recycling market for a heavy waste may be so far away that transportation costs exceed the recycling savings.
- The project may not generate enough waste in a particular category to interest an area recycler.
- A recycler may not have confidence the contractor will pay him for his services.

Obviously, the more recycling markets there are in the project area, the more likely the contractor is to find a market for his products. The other aid to his efforts is that the economic equation is tilting more and more in favor of recycling. Even when the contractor must pay a recycler to accept his waste, as well as pay for the container and hauling to transport it to him, the economics still tend to favor recycling. Fewer markets drive harder bargains, but a contractor who works hard at finding homes for his C&D waste should be able to accomplish his goals and save some money over the hypothetical cost of sending all of the waste to a landfill.

TYPES AND QUANTITIES OF WASTE

Identify and estimate all the various types and quantities of separate waste. Divorce the waste stream from the normal divisions of construction and review it from the market perspective. In this view, electrical, mechanical, and framing metal waste may all end up in one container because that is the way the recycler wants to receive it. He does not care about the construction schedule or phasing, when demolition occurs and when construction ends. He cares about when he can pick up a full container of scrap metal that meets his specifications. He particularly is concerned that the contractor not waste his time and money in calling for a pickup of a partial load, or one that is contaminated with products that he must cull out before processing.

As noted earlier, recycling is a bulk commodity enterprise. It operates on large quantities of low-margin, low-value items. A recycler makes his money by processing them for the end user as quickly and inexpensively as possible.

All of this points to the idea that, at least at this point in the recycling industry, it is a buyer's market. Recyclers provide a service to the contractor, and charge him for the privilege. The contractor must get in touch with this market as soon as he has estimates of his waste products, and must quickly understand various markets' needs and costs.

The contractor should start with the types of waste that are most prevalent on practically any site: concrete, gypsum board, wood, metals, and masonry. If he can find markets to accept these five types of waste, he has accomplished the heavy lifting of managing his waste management plan. On residential projects, these waste categories comprise approximately 75 percent of the total waste stream by weight. Nonresidential projects have a wide range of waste types, but on some renovation projects these five types of materials can account for as much as 90 percent of the total solid waste weight.

Plan the Work

Because contractors are responsible for much of the storage and hauling of recyclables on their sites, it makes sense for them to plan this work as efficiently as possible. Too many container rentals, or too many half-loads hauled to recyclers, are a waste of money and erodes the savings the contractor should be realizing from his recycling efforts.

The project manager and field superintendent should carefully plan the container and hauling schedule in conjunction with the project schedule. Match available site area and container storage to the volume of waste flowing from the demolition activities at a given

- Designate a plan leader (demolition subcontractor, field super-intendent, consultant, project manager).
- Educate and involve subcontractors and suppliers.
- Set up a recycling zone with sufficient space and clear signage.
- Police the program; prevent contamination.
- Monitor progress, report results, and reward participation.
- Create and collect documentation consistently.

Figure 6.1   Jobsite recycling tips.

time. Contractors with enough site area should not be shy about using overflow storage on pallets to handle short-term excesses and save the cost of renting an extra container.

Make sure contractors have an easy way to collect waste immediately adjacent to their work zone and transport it to the recycling zone. If site conditions permit, set containers adjacent to the area where they are working. Otherwise, use rolling hoppers or boxes that can be transported by front-end loaders or forklifts.

Linking the container and hauling schedule to the demolition/construction schedules is one of the most time-consuming, yet necessary, tasks facing the project manager and superintendent. As projects move forward from site clearing and demolition to new construction, the types and quantities of waste change, and so also must the characteristics of the recycling zone. Demolition waste, for instance, is bulky in nature and requires larger containers and overflow areas for the period in which it is being removed from the building. As demolition ends and the work moves into new construction, overflow areas may be less necessary, but more types of smaller containers may be required to make it easy for subcontractors to efficiently sort the waste. As the containers change, the manager of the recycling zone must also adjust the signage and communicate frequently with subcontractors to explain the changes and reinforce procedures to avoid contamination.

See Fig. 6.1 for a short list of jobsite recycling tips.

Educate Subcontractors and Suppliers

The waste management plan should include basic training provisions for subcontractors and suppliers involved in the project. As a practical matter, it will be very difficult to gather suppliers for an in-person training session, so the training for them should be in the form of a letter requesting their assistance in reducing the amount of packaging for their products. For suppliers who will be delivering a considerable amount of product to the jobsite, it may be worthwhile for the contractor to work with them one-on-one to reduce the packaging and material waste. A supplier of vinyl siding, for instance, may be amenable to allowing the subcontractor to pick up the product without packaging in an open-bed truck. There is one caveat with reducing supplier packaging. If a distributor simply removes the packaging from the manufacturer and disposes it of in his own Dumpster, waste is not being diverted from the landfill. The contractor needs some basic assurance (and in many cases it will be very basic) that the packaging that did not reach his site was recycled or was reused.

Unlike supplier training, subcontractor training on the operation of the plan should be mandatory. Following is a suggested list of methods to use in motivating subcontractors and suppliers to participate in the recycling plan:

■ Hold a plan orientation/kick off meeting.
■ Update plan progress in weekly jobsite meetings.
■ Encourage just-in-time deliveries.
■ Post targeted materials (signage).
■ Distribute tip sheets to job-site personnel.
■ Post goals/progress (signage).
■ Use formal agreements committing subcontractors to the program.
■ Require those who contaminate Dumpsters to re-sort.
■ Provide stickers, T-shirts, hats, or other incentives.
■ Publicly recognize participating subcontractors.
■ Take photos to document progress and share them.
■ At site visits with the owner, discuss waste management with job-site personnel.
■ Conduct periodic presentations for job-site personnel on waste issues.

Subcontractors are often not aware that their individual practices, outside of the plan, can greatly reduce the amount of solid waste generated on the site. Here is a list of suggested items for contractors to provide to subcontractors to consider, as part of educating them on their firm's impact:

■ Use less material.
■ Order in bulk.
■ Sell or donate salvaged materials.
■ Coordinate just-in-time deliveries.
■ Use precut and prefab components.
■ Reduce packaging wastes.
■ Use accurate materials estimating procedures.
■ Make use of scraps.
■ Avoid contaminating waste with toxic materials.
■ Plan to salvage.
■ Prevent damage to materials during handling.
■ Reduce the use of toxic materials.
■ Store materials properly.
■ Reuse salvaged materials.

Recycling Zone Management

The recycling zone on a construction site is an area shared by all the subcontractors and recyclers. The zone is managed, however, by the field superintendent, and he should make all parties aware of this fact. The cleanliness and orderliness of the recycling zone directly affects the contractor's waste management costs on a project. A sloppy, poorly maintained recycling area will inevitably lead to contaminated containers, delays in container drop-offs and pickups, and an overall poor attitude regarding the effectiveness of the plan. Subcontractors who see material strewn across the area and mixed waste on the ground will conclude, correctly one could argue, that the contractor really does not care about the details of managing the waste on his project, so why should they?

This attitude is fatal to a waste management program, and the solution to it is entirely within the contractor's control. His field superintendent must take full responsibility for policing the recycling zone and enforcing the rules against any subcontractors who stray from their responsibilities to neatly deposit and store waste. The superintendent needs to maintain clear signage, make sure the containers are emptied as needed, and assist contractors in neatly storing overflow waste when necessary. He must scold, remind, and fine any subcontractor who trashes his recycling zone, or otherwise does not comply with the obligations in his agreement with the contractor. This function is critical, and frankly, it is a role that only the field superintendent can fill.

See Fig. 6.2 for a summary of requirements necessary to satisfy a LEED prerequisite for operating a jobsite recycling center.

Following are some recycling zone management tips:

- Maintain a clear area with adequate access aisles and overflow storage.
- Keep the road base in good repair to avoid ponding and ruts.

 Compliance Connection

- Earn one point under LEED 2009 for Materials & Resources (MR) Credit 1.

- MR Credit 1 is prerequisite requirement, which must be achieved to earn LEED certification.

- MR Credit 1 awards *one point* for creating a dedicated jobsite area for the collection and storage of recycled materials.

- Materials collected in the recycling area must include: paper, corrugated cardboard, glass, plastics, and metals.

Figure 6.2   LEED recycling center prerequisite credit.

- Maintain clear signage for the containers at all times.
- Empty full containers promptly; set up any overflow on pallets.
- Keep wood, paper, and drywall waste containers covered.
- Maintain security for high-value waste.
- Maintain soil erosion and sedimentation control around the zone.
- Constantly police for contamination.
- Check rubbish containers routinely for misplaced recycling waste.

MAKE IT CONVENIENT

Use rolling hoppers, small Dumpsters, pickup trucks, front-end loaders, or any other device that will help subcontractors to follow the recycling plan. In general, the closer containers are to the work zone, the more likely the subcontractors are to recycle the material for which they are responsible. Rules are fine, but convenience and ease of use are even better. The dilemma facing the contractor is that he will want to place the recycling zone near an entrance to the site to enable ease of access for haulers and to limit the amount of aggregate base he must place to create a stabilized drive. This is reasonable, but if the recycling zone ends up far away from the work zone as a result, he must create some convenient way to encourage subcontractors to work efficiently yet still recycle.

PROMOTE THE RECYCLING PLAN

It is tempting to consider construction site tradesmen as hard-edged, jaded, coarse individuals with little interest in sustainable construction or reducing waste. Such a characterization is unfair to individuals who consider their work as fully skilled and professional (or more so) than that performed in an office or factory. Subcontractors and suppliers are like any other employees. They will respond to initiatives that capture their imagination and offer the inducement of personal or company recognition. Many are happy to heed the higher calling of environmental consciousness. Some have probably even felt pangs of guilt at walking past filled Dumpsters at the end of the workday, and wondered if more efficient practices would have reduced the amount of trash departing the site for the area landfill.

Contractors can encourage participation in the jobsite's recycling plan through a variety of strategies:

- *Subcontractor competition*: Each subcontractor will have a goal, which can be tracked through documentation managed by the contractor. Biweekly reports can track the compliance of subcontractors and rank them numerically. Biweekly prizes can be given to each member of the top-performing subcontractor crew.
- *Individual performance*: This is a discretionary prize awarded by the field superintendent for exceptional participation in the jobsite's recycling program. It may be awarded to a subcontractor employee who cleaned up a contaminated container (not his fault, of course), or to one who helped another subcontractor stack excess recycled material on pallets to prepare it for pickup.

■ *Miscellaneous awards*: The contractor can selectively recognize each subcontractor who is simply on-target with his recycling goals (not always an easy achievement), or alternatively recognize those that are most prompt or complete with their documentation or who provided the contractor with a lead on an unknown recycling market.

DOCUMENTATION

Contractors have two very good reasons to carefully document the weight and costs associated with waste management on their site: to document their compliance with municipal or contract requirements, and to track their costs to determine how the costs of the recycling waste management program compares with sending all waste to a landfill or incinerator. The first of these goals is the controlling factor, since a failure to comply with either a local ordinance or the provisions of the contract can have dire financial and legal consequences for the contractor. Of the two, the municipality may well be the more demanding in terms of insisting on independent documentation of C&D waste leaving the site. This is not to say that owners and architects will accept lesser documentation, only that a contractor who assumes that his records will be scrutinized by a municipal official (and who maintains them accordingly) will be on safer ground than one who assumes a friendly owner will give him a bye on the missing weight ticket.

In general, contractors should keep the following types of records to document their waste management efforts (also see Chap. 9 for detailed information on compliance documents):

1 Record (by weight and description) of all materials that leave the site, whether as landfill waste, salvage, or recycling.
2 Estimate of the weight and description of all materials that were kept and reused on site (photographs with scale elements are helpful).
3 Statements from distributors or manufacturers, including weight estimates, of the shipping materials they saved by using alternative protection on products they shipped to the jobsite.
4 Documentation of where these materials were delivered.
5 Documentation of what happened to the materials (the end use).
6 Documentation of the costs of hauling, recycling, and disposing of all wastes and recyclables that left the site.
7 Documentation of the costs of processing all materials that were reused on the site.

The Waste Management Plan should also state how the contractor intends to collect and manage this information, specifically, who is responsible for providing it from each recycler or subcontractor, and who is responsible for collecting it within the contractor's organization.

The quality and independence of documentation is key to the contractor meeting his goal of satisfying his recycling rate obligations. There will be times during the course of construction when he will not be able to obtain weight or end-use documentation for a container. That is the reality of running an enterprise involving a number of different players of varying capabilities. These instances should be few, however, because an inconsistent stream of contractor-generated documents undermines his credibility and raises serious questions as to the validity of his compliance. Following are the types of independent verification that should be satisfactory to either governmental or owner reviewers:

1 *Weight slips*: Each hauler or end market for site waste should provide a weight slip. Reliable recyclers know how essential the weight slips are to the process, and are accustomed to providing them. Weight slips should be obtained for each container that is hauled from the site.
2 *Certified scale weights*: Nonprofits, architectural salvagers, subcontractors, and other recycling markets will not always provide weight slips for loads they pick up. The contractor may find it very useful to contract with a local certified scale company and require any taker of waste products to have his weight documented. This is particularly important in cases in which the recycling rate is required by local ordinance and informal weight estimates may not be accepted as documented compliance.
3 *Documentation of recycling (or disposal):* Obtained from all end markets (in many cases, weight slips are adequate to provide this documentation).
4 *Transportation invoices*: Obtained from haulers or markets (in cases in which transportation is provided by the market).
5 *Recycling/disposal invoices/receipts*: Obtained from end markets.

The waste management plan should stipulate who is responsible for collecting and recording this information, where information will be stored, and who is responsible for preparing the compliance reports that will be issued periodically (typically monthly) to the owner or architect.

A few waste streams need special consideration. The waste management plan should include instruction on how documentation of the following wastes should be handled:

■ *Architectural salvage*: These types of items are not normally recorded by weight unless the contractor forces the issue. When the salvager escapes with the merchandise, the contractor's best case is to use the salvager's detailed proposal list of salvage items to assign reasonable weight estimates.
■ *Material recycled by subcontractors*: Many scrap metals have significant market value, and subcontractors may have assumed (or stipulated) they will handle the demolition waste under their contract. Typical subcontractors who may handle their own recycling include plumbers, electricians, HVAC contractors, and roofers. The contractor can certainly permit this, but he should insist on the full documentation from the subcontractor what he requires of recyclers, namely: description, weight slips, end-user documentation, and costs associated with the recycling.
■ *Equipment that is resold*: Items that are sold or donated are not often weighed. Such equipment may include: chillers, air conditioning or ventilation units, kitchen equipment, and industrial machinery. A certified scale requirement attached to any sale or donation is recommended, as is a detailed record of the size and model of the equipment, which can be helpful in acquiring a reasonable estimate of the weight. Again, a photograph is always helpful.

■ *Published weights*: Some equipment and building products may have published weights that can be documented with manufacturer's literature. Appurtenances, connections, and piping add weight to equipment, so the weight of a specific piece of equipment may not reflect the full weight of the recycled material. Nevertheless, when weight verification is lacking, this route may be all that is available as documentation to back a contractor's claim.

TRAINING

Training the contractor's and subcontractor's employees about the details of the recycling plan should take place prior to the beginning of work on the project. If all the subcontractors are identified at the beginning of the project (which is not always the case), the contractor can save some time by holding a mass training session for all subcontractors on a single day. It is asking much of a subcontractor, however, to request that he commit an entire crew on a workday to attend a one-hour training session for a specific project. The advantage of doing so, however, is that such a meeting establishes, in a very tangible way, that the recycling plan is important to the success of the project and that meeting the goals of the plan is a team responsibility.

The training session should have a short and succinct agenda:

1 *Goals of the program*: The recycling rate goal; the value and importance of recycling
2 *Materials to be recycled*: A quick review of all the materials to be recycled on the project
3 *Method of recycling*: The method(s) employed to achieve the goal; what they mean to the worker
4 *Sorting requirements*: Details of each type of material; what is not acceptable in each container; the problems caused by contamination
5 *Recycling zone rules*: Requirements for how the recycling zone will be managed
6 *Documentation*: Any subcontractor or submanaged supplier documentation that is required
7 *Tracking and reporting*: Incentives and prizes; reporting recycling progress to the jobsite workforce

Beyond the formal goals of the agenda, the training session must accomplish some less tangible, but equally important goals as well. Any endeavor involving a large group of people and various organizations must somehow engage the group. The participants must understand the importance of the recycling effort and accept their role in achieving the desired results. Here are three guiding principles for the contractor's recycling trainer to keep in mind as he presents the program:

1 *Convey enthusiasm*: This is not the easiest task with a group of hard-nosed subcontractors, but if the leader of the recycling effort is not enthusiastic about it, there is no hope the participants will be.

2 *Motivate the participants*: All the employees will be at the training session because they have been told, or compelled, to attend. This is not a prescription for healthy participation, so the trainer must motivate the attendees by explaining in clear terms why recycling is important, how it helps each firm and the planet, and how important each worker's role is to the success of the overall endeavor.

3 *Engage discussion*: Strong leadership is necessary to manage a jobsite recycling program, but not dictatorial management. The leader should encourage constructive criticisms and suggestions for how the program can be run more efficiently. In particular, the program manager wants to "draw out" misunderstandings or territorial issues that risk reducing compliance with program or creating contaminated loads.

SUBCONTRACTOR RECYCLING

Some contractors enter the project with the intent to recycle products under their trade. They may have included this assumption in their proposal to the general contractor, sometimes without his knowledge or consent. This occurs most often in projects that incorporate selective demolition carried out by the trades themselves, including work in which extensive electrical or plumbing work is being performed that will involve a fair amount of removal of existing material. These are the types of projects in which a demolition contractor would not normally be involved because the demolition work is trade-specific, and requires the judgment of the skilled tradesman as to how much must be removed and how much may be retained.

In these instances, it is reasonable for the individual subcontractor to make an assumption as to the amount of material that may be available for recycling, and to include that assumption under his contract. While the contractor who is surprised by this revelation after signing the contract with the subcontractor may be annoyed, he has probably not suffered any financial disadvantage as a result of the assumption. Every other skilled subcontractor has probably made the same assumption, and the contractor no doubt entered into a contract with the one who offered the lowest bid or otherwise offered the most value to the contractor. In sporting terms: *no harm, no foul*. The most important requirement for the contractor is that the weight and end-use of this material be documented and included in his overall recycling plan compliance figures. The subcontractor should be persuaded to provide this documentation to the contractor, and agree to do so as an addendum to his contract. While the weight value of this material, particularly in the electrical trade, may not be great, the products in question may be specifically listed in the specifications as subject to recycling.

Identify the Markets

Urban areas contain a well-developed network of recycling facilities, which can be located through directories published by the municipality, Web sites run by local non-profits, builder's exchanges, and contact with demolition subcontractors that are familiar with the area and the markets available for C&D waste. Municipalities in many cities have set up online waste exchanges, which contractors can use to identify potential markets for their waste.

The situation is more difficult in less developed areas, where recycling markets may be few and far away. In these areas, contractors need to rely heavily on local contacts (particularly demolition contractors) that may be aware of obscure markets for recycled goods that would otherwise escape his notice. These include area companies that will accept direct waste of a certain type, area nonprofit housing organizations, and national companies that may have regional distributors who accept waste on their behalf.

Contractors can also use recycling directories on the Internet (see the Resources section of this book) or local yellow page directories to identify companies that purchase recycled products. These companies are often listed under headings such as: scrap dealers, scrap metals, recycling centers, recycling services, waste reduction and recycling, architectural salvage, salvage, asphalt recycling, or reclamation.

Some demolition materials have little scrap value, but may be reinstalled as finish products. These may include vintage plumbing fixtures, cabinets, ornamental woodwork or decorative items, doors, and marble or granite slabs. For these types of products, investigate end users who may be willing to pay discounted prices, remove them at their cost, or in the case of nonprofit organizations, offer a tax deduction for the donation. Habitat for Humanity, for instance, is a nationwide organization that often accepts demolition products that are suitable for reuse in affordable homes they are building.

Calculate the Savings

See this book's Online Resources for a recycling plan savings calculator.

This calculation encapsulates the contractor's estimates of the total quantities of materials recycled and the total discarded. The recycling rate is simply the total quantity recycled divided by the sum of the quantity recycled plus the quantity thrown away.

When the plan is first developed, this will be an estimate, used to forecast an ultimate recycling rate and to assess changes in waste management procedures that will affect this rate. As the project moves along, it becomes a living record used to track progress toward recycling goals. If the rate runs below projections, the contractor should use the results documented in the plan to find out why, and (particularly if she needs a specific rate for LEED or other certification) use the plan to evaluate alternatives to increase the rate.

Problem Solving

Construction is rife with problems, and managing a recycling plan is no different. The best field superintendents and project managers in any organization are usually those who are the best problem solvers. Flexibility, resourcefulness, and a decent sense of irony are also useful in resolving the issues that arise in managing waste. See Fig. 6.3 for an example of a common site waste management problem—the overloaded container.

Figure 6.3   Overloaded container. © 2010, Joe Gough, BigStockPhoto.com.

Here is a list of hypothetical waste management problems and suggested solutions:

- *Problem*: The container is overflowing and the hauler is late.
    - *Solution*: Clear some area for overflow storage on pallets or a temporary wood platform until the hauler can empty the container.
- *Problem*: The container is in a subcontractor's way. He cannot work until it is moved.
    - *Solution*: Move the container or tell the contractor to work off of a lift.
- *Problem*: There is metal in the wood box and the driver refuses to take the container.
    - *Solution*: Reschedule the drive for the next day. Find the responsible subcontractor employee and make him clean it out.
- *Problem*: Someone keeps tossing their food waste into the mixed waste bin.
    - *Solution*: Remind all subcontractors of the requirements for mixed waste. Remind them also of the potential for fines in their agreement with the contractor.
- *Problem*: Area residents are using the project's Dumpsters for their personal waste.
    - *Solution*: Place a threatening sign at the recycling zone entrance. Lock Dumpsters, and tarp and lock roll-off containers at night.
- *Problem*: Rain is in the forecast and the hauler cannot pick up the cardboard roll-off. The recycler will not accept wet material.
    - *Solution*: Tarp the roll-off container and vow to only place cardboard in large Dumpsters from now on.
- *Problem*: The shingles are full of nails and it will take too long to remove them. Can they be put in the shingle container?
    - *Solution*: Check the requirements of the shingle recycler to see if he accepts nails as part of the load. If he doesn't, consider renting a grinder for use on-site, and make sure it can separate metals from the shingle waste.

- *Problem*: The monthly prizes were awarded to subcontractors and one worker was accidentally left off the list.
  - *Solution*: Put out a special supplemental bulletin, and award him his prize at the next weekly subcontractor meeting.
- *Problem*: Recycling will take too long. We won't finish on schedule.
  - *Solution*: Sorting of materials, once mastered, should take no longer than mixed waste disposal. All subcontractor waste has to go in a container, anyway. Small rolling hoppers can be placed near the work zone.
- *Problem*: There is not enough room on the site to recycle. All those containers take up too much space.
  - *Solution*: Smaller recycling Dumpsters, located strategically around the site if necessary, can take the place of large roll-offs for mixed waste.
- *Problem*: My workers can't remember how to sort the material. This is too complicated for them.
  - *Solution*: Have clear signage on the containers and give clear instructions to the workers. Laminated cards for each subcontractor employee (with pictures) can help.
- *Problem*: Recycling is too expensive. I didn't include it in my bid.
  - *Solution*: It is a part of the contractor/subcontractor agreement the subcontractor signed, and includes penalties if he does not comply. Besides, it will save him money.
- *Problem*: My workers will have to work overtime to sort the recycling. I'll have to pay them and send you a change order.
  - *Solution*: Waste management is part of the subcontractor obligation under his agreement. No change orders for the purpose of recycling will be considered.

Tips for Effective Management

If there is only one critical requirement in running a C&D waste management operation, it is that at least two people in the contractor's employ have full knowledge and responsibility for managing the entire plan. This does not mean that one individual must do everything, but this individual must have participated in every step of the plan, from the original site assessment to negotiating contracts with the recyclers, and have made the hauling and container arrangements.

Typically, this person will be the contractor's project manager. He is normally responsible for all business arrangements relating to the project, and he thus will be the savviest at knowing the contract conditions, exclusions, and pitfalls of each party. The project manager, therefore, is normally the planner and overseer of the entire waste management operation on the site. He cannot manage the day-to-day operations of the recycling effort, however, without the help of the field superintendent. It would be easy enough to claim that the field superintendent is the "operations end" and the project manager is the "management side" of a waste management plan. The reality, however, is that the field superintendent needs to possess almost the full range of information that the

Figure 6.4   Effective recycling management means pairing equipment, labor, and schedule. *Courtesy of Tristan Winkler.*

project manager holds. The field superintendent, for instance, must be fully aware of the recycler's conditions on what the waste loads must, and must not, contain. This person must also be aware of any special arrangements the project manager has made with subcontractors regarding their own handling of waste under their trades. If office personnel in the contractor's main office are collecting documentation, the field superintendent needs to be in regular communication regarding which haulers are failing to, for example, stop by the certified scales on the way to the recycling facility. The field superintendent can be essential in reminding the haulers at the time of pickup that this is an obligation of their contract. (See Fig. 6.4).

Waste Management Plan Checklist

The waste management plan should be the contractor's internal working document, containing the nuts and bolts of how he will manage the on-site recycling effort. It is an organic plan, shifting as the needs and realities of the waste management stream dictate. Although the plan may be shared with the owner and architect, the contractor should be very clear in communicating that this checklist, separate from the compliance document he owes to the owner (see an example of a Waste Management Compliance Document later in this chapter), is strictly for his management use. A draft Waste Management checklist should contain the following minimum information:

1 Name the individual(s) responsible for waste prevention and management.
2 List the actions that will be taken to reduce solid waste generation.
3 Schedule the regular meetings to address waste management.
4 Describe the specific approaches to be used in recycling/reuse.
5 List the waste characterization, including estimated material types and quantities.
6 Calculate local landfill estimated costs, assuming no salvage or recycling.
7 Identify local and regional reuse programs.
8 List the specific waste materials to be salvaged and recycled.
9 Estimate the percentage of waste diverted by this plan.
10 Identify recycling facilities to be used.
11 Identify materials that cannot be recycled or reused.
12 Describe the means by which any materials to be recycled or salvaged will be protected from contamination.
13 Describe the means of collection and transportation of the recycled and salvaged materials.
14 Calculate anticipated net cost or savings.
15 Require the subcontractors to document their actual waste diversion performance throughout the project. The waste management plan should also include progress reporting procedures to record actual diversion and cost corresponding to each diversion and cost estimate.
16 Incorporate the plan into the contractor's quality control and owner's quality assurance processes.
17 Establish that the responsibilities and costs associated with the construction and demolition waste are the responsibility of the contractor, and that the contractor is allowed to accrue the economic benefits of reuse or recycling. These include cost avoidance through reduced debris tipping expenses, revenues from salvaged and recycled materials, and cost savings by reusing demolition materials on the jobsite.
18 Evaluate the project's waste reduction potential.
19 Identify and implement efficient construction techniques.
20 Strategize to determine methods to reduce, reuse, and recycle.
21 Identify target materials for recycling.
22 Identify local alternative waste disposal options (i.e., recyclers, salvagers).
23 Compare rates and other factors and negotiate services.
24 Finalize agreement.
25 Finalize target material selection.
26 Set waste reduction goals; calculate recycling (diversion) rate.
27 Develop a communications and training plan.
28 Develop a waste management motivational plan.
29 Develop a waste management evaluation plan.
30 Ensure subcontractor commitment.
31 Assign individuals responsible for the plan within the contractor's forces, and with each subcontractor.

32 Establish contacts with local recyclers, salvagers.

33 Plan an educational program to foster workforce acceptance.

Draft Waste Management Compliance Document

Unlike the waste management plan, the *Waste Management Compliance Document* is the formal document the contractor uses to demonstrate compliance with the terms of the contract for construction. If the architect has stipulated the minimum elements of a waste management plan in the specifications, the contractor should hew fairly close to these minimum requirements. Why not use the *waste management plan* for both purposes? Because the plan will contain cost, subcontractor, and other information that will change as the project moves forward.

Indeed, the intent of the waste management plan is that it will change to adapt to circumstances, whereas the compliance document will always reflect the same estimated quantities and goals. The end results, of course, will be what they are, and the compliance document will be assessed in relation to the requirements the owner, architect, or municipality placed on the contractor. The waste management plan allows the contractor the freedom to plan beyond those requirements.

Unless the owner has specifically demanded cost information about the waste management efforts as part of the contract, he is not entitled to see them. The owner typically intends to saddle the contractor with all costs associated with waste management on the project. As this is the case, the contractor is not under any obligation to share the savings achieved through sustainable waste management practices unless his agreement with the owner obligates him to do so. He may choose to share this information in the spirit of helping the owner to publicize his recycling successes, but he should make clear that the waste management savings are the contractor's to keep.

See this book's Online Resources for the full form in an editable format.

WASTE MANAGEMENT COMPLIANCE DOCUMENT GUIDELINES

1 This project shall achieve a construction and demolition recycling rate (also referred to as the diversion rate) of *50 percent (insert actual required percentage here)*. This represents the percentage of total construction waste diverted from the landfill through reduction, reuse, and recycling.

2 Work forces shall generate the least amount of waste possible by planning and ordering carefully, following all proper storage and handling procedures to reduce waste, and reusing materials wherever possible. Waste materials generated shall be salvaged for donation or resale, or separated for recycling to the extent that is economically feasible.

3 The contractor shall be allowed to count any waste reduction achieved through reduced packaging, bulk orders, or other means, as long as such waste is properly documented.

4 The waste management chart (see section Waste Management Chart) identifies the demolition and new construction waste materials expected to be generated on this project, as well as the estimated quantities for the disposal method(s) for each material.

5 Each contractor and subcontractor will be provided with a copy of the waste management plan and trained in its use. Each subcontractor will be responsible for ensuring his employees comply with the plan.

6 Waste management activities will be reviewed at each project meeting, and a progress report will be issued monthly to the owner and architect in conjunction with the periodic application for payment.

7 A secure and managed recycling zone shall be created on the jobsite. All containers will be clearly labeled and lists of accepted/unaccepted materials will be posted throughout the site and provided to each subcontractor.

WASTE MANAGEMENT CHART

Demolition Material   Estimated quantity and total weight reuse/recycle/dispose

- Asphalt
- Cardboard and paper
- Concrete
- Glass
- Gypsum board
- Hazardous waste
- Insulation and packing
- Masonry (brick and block)
- Metal (ferrous)
- Metal (nonferrous)
- Miscellaneous
- Plastics and plastic fabrications
- Porcelain fixtures
- Roofing shingles or membranes
- Site residuals
- Wood and wood products

See Tables 6.1 and 6.2 for charts of volume/weight conversions for common materials. See Chap. 12 for more extensive conversion lists.

Summary

- The recycling waste management plan has a number of purposes, with the final objective being to determine the recycling rate.
- An on-site assessment should be conducted to determine the types and estimated quantities of recycled waste products.

TABLE 6.1   C&D WASTE VOLUME/WEIGHT CONVERSIONS

MATERIAL	POUND/CUBIC YARD	YARDS/TON
Mixed construction waste	350	5.7
Wood	300	6.7
Drywall	500	4.0
Rubble	1,400	1.4
Cardboard	100	20.0
Landscape debris	240–400	5–8

TABLE 6.2   C&D WASTE SI VOLUME/WEIGHT CONVERSIONS

KILOGRAMS PER MATERIAL	CUBIC METERS PER CUBIC METER	METRIC TONNE
Mixed construction waste	208	4.8
Wood	178	5.6
Drywall	297	3.4
Rubble	831	1.2
Cardboard	59	16.9
Landscape debris	142–237	7.0–4.2

■ Planning the work includes tying the demolition and construction schedule to container rentals and pickups.

■ Effective recycling zone management includes: adequate clearances, cleanliness, clear signage, and adequate overflow areas.

■ Subcontractor training is necessary to communicate the details of the site recycling plan and to establish team spirit in accomplishing it.

■ Establishing clear standards and responsibilities for waste documentation is key to achieving a verifiable recycling rate.

■ The waste management plan is the contractor's internal working document.

■ The waste management compliance document is the formal document the contractor uses to demonstrate compliance with her contract.

COMPLIANCE CONNECTION

Leadership in Energy and Environmental Design (LEED)® is a third-party certification program and a widely accepted benchmark for the design, construction, and operation of sustainable buildings (Fig. 7.1). Developed by the U.S. Green Building Council (USGBC) in 1998 through a committee process involving a wide range of nonprofit, industry, and governmental groups, LEED serves as a design and construction template for sustainable buildings of all types and sizes. As of 2009, USGBC estimates that more than 4.5 billion ft² (4.18 billion m²) of building area have been designed using the LEED program as guidance.[1]

The certification process, administered by a USGBC spinoff organization called the Green Building Certification Institute (GBCI), determines the appropriateness of buildings for LEED certification through an online submission process, using templates for each of the seven types of construction recognized under the LEED system. LEED certification is available for all of the following building types: general new construction, major renovation, existing buildings, commercial interiors, core and shell, schools, and homes. As of this writing, LEED certification for retail space, neighborhood development, and health-care facilities are currently in development.

LEED New Construction (LEED NC) is the category most often experienced by general contractors working on commercial projects. This category covers, as the name implies, almost all new construction work (see the core and shell exception below), and major renovation work as well. USGBC defines major renovation as: "A major renovation involves major HVAC renovation, significant envelope modifications, and major interior rehabilitation." In cases with lesser degrees of renovation not meeting the criteria for LEED NC, owners and contractors should use the LEED for Existing Buildings: Operation and Maintenance criteria (LEED EB).

To qualify for certification, projects must meet certain prerequisites and earn additional performance points in six base categories of sustainable design. The six categories include: (1) Sustainable Sites (SS), (2) Water Efficiency (WE), (3) Energy and Atmosphere (EA), (4) Materials and Resources (MR), (5) Indoor Environmental Quality (IEQ), and (6) Regional Priority (RP). An additional category, Innovation in Design (ID), addresses unusual or exceptional situations, as well as design situations not covered under the six environmental categories.

LEED

Leadership in Energy and Environmental Design (LEED) is a program of the United States Green Building Council, a voluntary organization of building industry professionals, companies, environmental organizations, and governmental groups.

LEED is a certification standard for buildings, based on points granted for meeting sustainability goals in seven different categories.

Figure 7.1   LEED description.

The USGBC describes the importance of the rating system in their LEED 2009 update:

In LEED 2009, the allocation of points between credits is based on the potential environmental impacts and human benefits of each credit with respect to a set of impact categories. The impacts are defined as the environmental or human effect of the design, construction, operation, and maintenance of the building, such as greenhouse gas emissions, fossil fuel use, toxins and carcinogens, air and water pollutants, indoor environmental conditions.

LEED projects can earn up to 110 base points, including six *Innovation and Design Process* credits and up to four *Regional Priority* points for obtaining products within a 500-mi (805-km) radius of the construction site. Projects are recognized in four progressive categories, according to the number of points they earn:

- Certified project: 40 to 49 points
- Silver project: 50 to 59 points
- Gold project: 60 to 79 points
- Platinum project: 80 points and above

The USGBC New Construction (NC) standards state that their standards may apply to new buildings or to renovation projects where a significant portion of the facility is undergoing renovation. The LEED program does make a distinction, however, regarding projects that are designed and constructed to be partially occupied by the owner or developer, with the rest occupied by tenants. In these projects, USGBC claims the owner or developer has direct influence over the portion of the building they occupy, and to pursue certification under the LEED NC category the owner must therefore occupy 50 percent or more of the building's leasable area. Where this is not the case, the project is not suitable for LEED NC credits and the owner should pursue LEED for Core & Shell certification.

USGBC has introduced a separate system for LEED certification of homes. This system offers a total of 136 points, 45 of which are the minimum required for LEED home certification. Curiously, only three of the points can be earned for jobsite waste management during construction (though an additional nine points can be earned for reducing framing waste in material orders and the use of off-site fabrication for structural components).

Total LEED program-related costs vary by the project type and size. Registration of a project with GBCI costs approximately $450 (for USGBC members) and $600 (for nonmembers). LEED certification costs vary with project size, but USGBC states the average cost is around $2000. Indirect costs may include independent LEED consultants and building commissioning. These costs vary dramatically, based on the size and complexity of the project.

Waste Management Credits Under LEED[2]

LEED classifies waste management credits under the *Materials and Resources* (MR) section.

Waste management practices are also described in this section, under which two points are offered for properly managing construction and demolition waste on projects. One point is awarded to projects that successfully divert 50 percent of the total waste generated from construction or demolition. A second point is awarded for diverting 75 percent of the total waste. The quantity of total waste cannot include any hazardous materials. The intent of this credit is to divert construction and demolition waste from landfill disposal and return resources back to the manufacturing process. See Fig. 7.2.

 Compliance Connection

■ Materials and Resources (MR) Prerequisite 1: Storage and Collection of Recyclables

■ MR Credit 1.1: Building Reuse—Maintain Existing Walls, Floors and Roof

■ MR Credit 1.2: Building Reuse—Maintain Existing Interior Nonstructural Elements

■ MR Credit 2: Construction Waste Management

■ MR Credit 3: Materials Reuse

■ MR Credit 4: Recycled Content

Figure 7.2   **LEED waste management and recycling credits.**

In order to earn these credits, the contractor must complete a template form on the USGBC Web site. This template closely mirrors a conventional waste management plan, in that it requires a listing of waste materials by weight, and information about whether they will be sent to landfills or diverted from them. Calculations can be done by weight or volume, but must be consistent throughout the template.

A narrative description is also required, along with questions that prompt the contractor to analyze the project, set up a recycling zone for the site, investigate waste recyclability, train and motivate subcontractors and workers, and keep pertinent documentation.

MR PREREQUISITE 1: STORAGE AND COLLECTION OF RECYCLABLES

The LEED system requires prerequisites in each category. A prerequisite must be satisfied before any other credits can be earned. In the Materials and Resources category of LEED, the prerequisite requires that the contractor create a jobsite area for collecting and storing recycled material. The intent of this requirement is clearly to encourage contractors to reduce the amount of material diverted to landfills through a construction recycling program. Oddly, the requirement does not stipulate that a recycling program be conducted in conjunction with the on-site requirement, though earning later credits in the Materials and Resources category would necessarily require the contractor to create such a program.

Documentation Requirements   Provide a dedicated area (or multiple areas) on the construction site for collecting and storing recycled materials related to the demolition or construction of the building. LEED stipulates that the following materials must be recycled and stored in the recycling area: paper, corrugated cardboard, glass, plastics, and metals.

Strategies to Meet the Requirements

- Create a recycling area that is appropriately sized for the anticipated amounts of recycled material resulting from the demolition or construction.
- Coordinate container sizes and hauler pickups with the construction schedule to minimize the area required for the recycling zone.
- Consider using equipment such as cardboard balers, aluminum can crushers, recycling chutes, and smaller collection bins at work areas to reduce the amount of space required for on-site storage.

MR CREDIT 1.1: BUILDING REUSE: MAINTAIN EXISTING WALLS, FLOORS, AND ROOF (ONE TO THREE POINTS)

The intent of this credit is to extend the life of existing buildings, conserve natural resources, retain valuable cultural resources, reduce construction and demolition waste, and reduce the environmental effects of new buildings as they relate to materials manufacturing and transport.

Requirements

- Maintain a minimum percentage (based on surface area) of existing building structure (including structural floor and roof decking) and envelope (exterior skin and framing, excluding window assemblies and nonstructural roofing material).
- Hazardous materials that are remediated as a part of the project scope shall be excluded from the calculation of the percentage maintained. If the project includes an addition to an existing building, this credit is not applicable if the square footage of the addition is more than two times the square footage of the existing building.
- One point is earned for maintaining 55 percent of the required building components.
- Two points are earned for maintaining 75 percent of the required building components.
- Three points are earned for maintaining 95 percent of the required building components.

Strategies to Meet the Requirements

- Consider reuse of existing buildings, including the structure, envelope, and elements.
- Remove hazardous elements that pose a contamination risk to building occupants, while upgrading components that will improve energy and water efficiency, such as windows, mechanical systems, and plumbing fixtures.

MR CREDIT 1.2: BUILDING REUSE: MAINTAIN 50 PERCENT OF INTERIOR NONSTRUCTURAL ELEMENTS (ONE POINT)

Intent   Same as for MR Credit 1.1.

Requirements

- Use existing interior nonstructural elements (interior walls, doors, floor coverings, and ceiling systems) in at least 50 percent (by area) of the completed building (including additions).
- If the project includes an addition to an existing building, this credit is not applicable if the square footage of the addition is more than two times the square footage of the existing building.

Strategies to Meet the Requirements

- Consider reuse of existing buildings, including structure, envelope, and interior nonstructural elements.

■ Remove elements that pose contamination risk to building occupants and upgrade components that would improve energy and water efficiency, such as mechanical systems and plumbing fixtures.

■ Quantify the extent of building reuse.

MR CREDIT 2: CONSTRUCTION WASTE MANAGEMENT (ONE TO TWO POINTS)

The intent of this credit is to divert construction, demolition, and land-clearing debris from disposal in landfills and incinerators, redirect recyclable recovered resources back to the manufacturing process, and redirect reusable materials to appropriate sites.

Requirements

■ Recycle and/or salvage a minimum percentage of nonhazardous construction and demolition debris.

■ Develop and implement a construction waste management plan that, at a minimum, identifies the materials to be diverted from disposal and whether the materials will be sorted on-site or commingled.

■ Excavated soil and land-clearing debris do not contribute to this credit. Calculations can be done by weight or volume, but must be consistent throughout.

■ Earn one point by recycling or salvaging at least 50 percent of the waste material from a construction site.

■ Earn two points by recycling or salvaging at least 75 percent of the waste material from a construction site.

Strategies to Meet the Requirements

■ Establish goals for diversion from disposal in landfills and incinerators and adopt a construction waste management plan to achieve these goals.

■ Consider recycling cardboard, metal, brick, acoustical tile, concrete, plastic, clean wood, glass, gypsum wallboard, carpet, and insulation. Designate a specific area or areas on the construction site for segregated or commingled collection of recyclable materials, and track recycling efforts throughout the construction process.

■ Identify construction haulers and recyclers to handle the designated materials. Note that diversion may include donation of materials to charitable organizations and salvage of materials on the site.

Documentation Requirements   Each credit a project attempts to achieve for LEED certification requires documentation to prove the activity was completed in accordance with the requirements.

LEED uses customized online fillable form templates for projects to certify that LEED requirements are met for each prerequisite and credit. Additional documentation may still be required. Because the contractor is responsible for construction waste management, the contractor will generally be responsible for completing the required forms. The person responsible for documenting compliance for certification will need to prepare the LEED Letter Template, signed by the architect, owner, or other responsible party, tabulating the total waste material, quantities diverted from landfills, and the means through which they were diverted.

A portion of the credits in each application will be audited, and the contractor should be prepared with backup documentation for credits related to jobsite waste management. The following documentation is recommended as good practice for operating an environmentally responsible jobsite, and as backup for an audit of your project's LEED application.

LEED audits will generally require a waste management plan and regular submittals tracking progress. Contractors should develop a plan that shows how they will achieve the required recycling rate, including materials to be recycled or salvaged, cost estimates comparing recycling costs to disposal fees, materials handling requirements, and how the contractor will train his employees and subcontractors to comply with the plan's requirements.

MR CREDIT 3: MATERIALS REUSE (ONE TO TWO POINTS)

The intent of this section is to reuse building materials and products in order to reduce demand for virgin materials and to reduce waste, thereby reducing the effects associated with the extraction and processing of virgin resources.

Requirements

- Use salvaged, refurbished, or reused materials such that the sum of these materials constitutes a minimum percentage, based on cost, of the total value of materials for the project.
- Mechanical, electrical, and plumbing components and specialty items such as elevators and equipment shall not be included in this calculation. Only include materials permanently installed in the project.
- Furniture may be included.
- Earn one point by reusing 5 percent of the total value of materials on a project.
- Earn two points by reusing 10 percent of the total value of materials on a project.

Strategies to Meet the Requirements

- Identify opportunities to incorporate salvaged materials into building design and research potential material suppliers.
- Consider salvaged materials such as beams and posts, flooring, paneling, doors and frames, cabinetry and furniture, brick, and decorative items.

MR CREDIT 4: RECYCLED CONTENT (ONE TO TWO POINTS)

The intent of this credit is to increase demand for building products that incorporate recycled-content materials.

Requirements

- Use materials with recycled content such that the sum of postconsumer recycled content plus half of the preconsumer content constitutes a minimum percentage (based on cost) of the total value of the materials.
- Recycled-content value of a material assembly shall be determined by weight. The recycled fraction of the assembly is then multiplied by the cost of assembly to determine the recycled-content value.
- Mechanical, electrical, and plumbing components and specialty items such as elevators shall not be included in this calculation.
- Only include materials permanently installed in the project.
- Earn one credit for recycling 10 percent of the total value of the project.
- Earn two credits for recycling 20 percent of the total value of the project.

Strategies to Meet the Requirements

- Establish a project goal for recycled-content materials and identify material suppliers that can achieve this goal.
- During construction, ensure that the specified recycled-content materials are installed.
- Consider a range of environmental, economic, and performance attributes when selecting products and materials.

MR CREDIT 5: REGIONAL MATERIALS: EXTRACTED, PROCESSED, AND MANUFACTURED REGIONALLY (ONE TO TWO POINTS)

The intent of this credit is to increase demand for building materials and products that are extracted and manufactured within the region, to support the use of indigenous resources, and to reduce the environmental effects resulting from transportation.

Requirements

- Use building materials or products that have been extracted, harvested, or recovered, as well as manufactured, within 500 mi (805 km) of the project site for a minimum percentage (based on cost) of the total materials value. If only a fraction of a product or material is extracted/harvested/recovered and manufactured locally, then only that percentage (by weight) shall contribute to the regional value.
- Mechanical, electrical, and plumbing components and specialty items such as elevators and equipment shall not be included in this calculation. Only include materials permanently installed in the project. Furniture may be included, providing it is included consistently in MR Credit 3.

- Earn one point for using 10 percent regional materials.
- Earn two points for using 20 percent regional materials.

Strategies to meet the requirements

- Establish a project goal for locally sourced materials, and identify materials and material suppliers that can achieve this goal.
- During construction, ensure that the specified local materials are installed and quantify the total percentage of local materials installed.
- Consider a range of environmental, economic, and performance attributes when selecting products and materials.

Innovation and Design Process Credits

LEED offers credits under the *Innovation and Design Process* category for sustainable practices in areas not addressed in one of the five standard categories, or for exceptional performance in some aspect of sustainable design that has the potential to be duplicated in other projects. Of the published LEED interpretations of granted innovation credits, none has so far been awarded for construction waste management practices. At least one innovation credit has been awarded for voluntarily moving an entire building on the same site in order to preserve it. Though an exemplary reuse of existing materials, this was clearly an exceptional circumstance that cannot be replicated on many projects. The fact that innovation credits have not been awarded for innovative recycling efforts does not mean that such a credit is impossible, though with standard recycling rates on commercial projects routinely exceeding 50 percent, it would take a truly high percentage to qualify as a potential innovation. And according to USGBC guidelines for their innovation credit, the technique used to achieve that very high percentage would need to be such that it could be duplicated on other projects.

Waste Management Policy Compliance

The field superintendent should keep a log to track loads of any waste material leaving the job site. Each load of recycled material leaving the site should be tracked on a load tracking form. The form will identify the following:

1 The type of material (mixed metals, wood scrap, etc.)
2 The date
3 The hauler
4 The destination (market or landfill)
5 The weight
6 The documentation source

See this book's Online Resources for a sample form. Along with the log to track loads, the field superintendent needs to maintain a separate log for recycled materials processed on the jobsite and reused in another application. Examples of uses of such materials could be:

- Concrete or masonry waste used as aggregate base for roadways or parking
- Vegetative waste used as mulch
- Structural steel or joists kept in place, removed and replaced, or reworked into new construction
- Dimensioned framing or heavy timber kept in place or reused in new construction

For those projects seeking LEED certification, other specific documentation must be maintained. See the beginning of this chapter for information on the specific LEED credits related to C&D waste, and see Chap. 8 for other types of certification programs and their requirements.

References

1. "About USGBC." U.S. Green Building Council (USGBC). November 29, 2009 <http://www.usgbc.org/DisplayPage.aspx?CMSPageID=124>.
2. "LEED 2009." United States Green Building Council (USGBC). November 19, 2009 <http://www.usgbc.org/ShowFile.aspx?DocumentID=5546>.

8

OTHER GREEN CERTIFICATION

AND CODE PROGRAMS

Although the U.S. Green Building Council's LEED program is the most well known, other programs have been developed by national organizations to promote sustainable construction. These programs, some of which include certification components, all incorporate aspects of recycling construction and demolition waste to varying extents (see Fig. 8.1).

International Code Council Evaluation Service (ICC-ES) Sustainable Attributes Verification and Evaluation™ (SAVE™) Program

ICC-ES created the SAVE program to satisfy a critical need in the world of green building: the need to verify manufacturers' claims regarding the sustainable attributes of their products. With input from major industry partners, ICC-ES developed guidelines that examines the attributes of a product or system, beginning with raw material acquisition and progressing through final manufacturing and packaging (the "cradle-to-gate" concept). Companies may have their products, materials, and/or systems evaluated within the scope of one or more of nine categories. These categories are[1]:

- EG101: Evaluation Guideline for Determination of Recycled Content of Materials
- EG102: Evaluation Guideline for Determination of Biobased Material Content
- EG103: Evaluation Guideline for Determination of Solar Reflectance, Thermal Emittance, and Solar Reflective Index of Roof Covering Materials
- EG104: Evaluation Guideline for Determination of Regionally Extracted, Harvested, or Manufactured Materials or Products

Compliance Connection

■ International Code Council's ICC-ES SAVE™ program

■ International Code Council's International Green Construction Code (IGCC)

■ National Association of Homebuilders (NAHB) Green Building Program

■ U.S. Environmental Protection Agency Energy Star Program

■ National Institute of Building Sciences (NIBS) Whole Building Design Guide

■ Green Building Initiative Green Globes® Program

■ Florida Green Home and Commercial Buildings Standards

■ GreenPoint Rated New Home Program (California)

■ United Kingdom: BREEAM

■ Australia: Green Star

Figure 8.1   **Other green certification and compliance programs.**

■ EG105: Evaluation Guideline for Determination of Volatile Organic Compound (VOC) Content and Emissions of Adhesives and Sealants
■ EG106: Evaluation Guideline for Determination of Volatile Organic Compound (VOC) Content and Emissions of Paints and Coatings
■ EG107: Evaluation Guideline for Determination of Volatile Organic Compound (VOC) Content and Emissions of Floor Covering Products
■ EG108: Evaluation Guideline for Determination of Formaldehyde Emissions of Composite Wood and Engineered Wood Products
■ EG109: Evaluation Guideline for Determination of Certified Wood and Certified Wood Content In Products

Once the evaluation is complete, a *Verification of Attributes Report*™ (VAR™) is generated, which provides technically accurate product information that is beneficial to owners and contractors seeking to qualify for points under national and international green rating systems [including the U.S. Green Building Council's LEED program, Green Building Initiative's (GBI) Green Globes, or the proposed ICC *National Green Building Standard*™]. VAR reports are only prepared for those products that are verified by ICC-ES as having the product attributes and meeting the manufacturer's environmental claims. To continue to claim certification under the program, manufacturers must allow annual inspections by ICC-ES at the product's manufacturing site to ensure that the attributes of the product described in the VAR remain true. Additional information about the ICC-ES SAVE program is available at http://www.icc-es.org/.

International Code Council (ICC) International Green Construction Code (IGCC)

The ICC, in partnership with the American Institute of Architects (AIA) and the American Society for Testing and Materials (ASTM) is working to develop a new set of green codes under a multiyear initiative called "International Green Construction Code: Safe and Sustainable by the Book."[2] This initiative will include collaboration among a wide range of professional, trade, green building leaders, as well as outreach and feedback from our members and the general public. The resulting code will cover all aspects of sustainability in the built environment, including drawing from existing codes and standards to create a single universal code. The proposed code will apply to new construction and renovations, and will link in closely with existing ASTM standards. Slated for implementation in 2010, the IGCC will, for the first time, provide a regulatory framework to assist municipalities and code officials in understanding and administering green construction. Additional information regarding the IGCC is available at http://www.iccsafe.org/.

National Association of Homebuilders (NAHB) Green Building Program

The National Association of Homebuilders developed in 2006 a document called *The Green Homebuilding Guidelines*. Created with the involvement of an advisory committee representing more than 90 national nonprofit affordability, solar, and environmental organizations, as well as some of the largest homebuilders in the nation, The Green Homebuilding Guidelines are intended for builders "engaged in or interested in green building products and practices for residential design, development, and construction." See Table 8.1 for a points summary.

The guidelines cover the following aspects of residential construction[3]:

1 *Lot design, preparation, and development*: Selecting and developing a site
2 *Resource efficiency*: Using sustainable materials and reducing waste during construction
3 *Energy efficiency*: Designing more energy-efficient housing; using solar principles
4 *Water efficiency*: Using water resources more efficiently
5 *Indoor environmental quality*: Reducing indoor pollutants and enhancing air quality
6 *Operation, maintenance, and homeowner education*: Using better home maintenance and operating more efficiently
7 *Global impact*: Purchasing materials from sustainable manufacturers; using paints containing a low level of volatile organic compounds (known also as low-VOC paints)

TABLE 8.1   NATIONAL ASSOCIATION OF HOMEBUILDERS GREEN BUILDING PROGRAM POINTS

	BRONZE	SILVER	GOLD
Lot design, preparation, and development	8	10	12
Resource efficiency	44	60	77
Energy efficiency	37	62	100
Water efficiency	6	13	19
Indoor environmental quality	32	54	72
Operation, maintenance, and homeowner education	7	7	9
Global impact	3	5	6
Additional points from sections of your choice	100	100	100

The guidelines are comprehensive in nature, covering the full spectrum of housing development and offering practical, if general, guidance on how to develop more energy-efficient, sustainable housing. Within the guidelines, NAHB sets up a basic rating system—bronze, silver, and gold—that builders can achieve through meeting certain criteria. The NAHB guidelines address construction waste issues in two subsections, 2.3 and 2.5, of the *Resource Efficiency* section. Following is a summary of the waste management requirements included in the NAHB guidelines.[2]

- *Section 2.3.1*: Disassemble existing buildings (use deconstruction) instead of demolishing.
- *Section 2.3.2*: Reuse salvaged materials as much as possible. Document compliance with a list of components that were reused in the project.
- *Section 2.3.3*: Dedicate a recycling area and provide on-site containers for waste management, such as the sorting and reuse of scrap building materials. Document compliance with a copy of the project's C&D waste management plan.
- *Section 2.5.1*: Develop and implement a construction and demolition program. Document compliance with a copy of C&D waste management plan.
- *Section 2.5.2*: Conduct on-site recycling efforts, such as using a site grinder, applying waste materials on-site, or reducing transportation-related costs for material deliveries. Document compliance with a copy of the C&D waste management plan that is posted at the jobsite.
- *Section 2.5.3*: Recycle construction waste off-site, including waste such as wood, cardboard, metals, drywall, plastics, roofing shingles, concrete, and masonry. Document compliance using contractual agreement between the recycling firm and the builder, documentation of materials that have been recycled, or a list of components that were recycled.

This system is not related in any way to the USGBC LEED rating system. Indeed, no representative of USGBC was listed as a member of the advisory group, and the guidelines do not refer to the LEED system at any point. The criteria are met through a system of varying types of certification (dependent on the particular credit area), ranging from builder or installer certification, local building authority statement, calculations, or documentation provided by a design professional.

More information on the National Homebuilders Association Green Building Program is available at www.nahb.org.

U.S. Environmental Protection Agency Energy Star Program

The Energy Star program developed by the U.S. Environmental Protection Agency (EPA) is a national energy performance rating system that benchmarks the energy performance of a wide range of commercial facilities relative to the performance of similar facilities across the United States. To be eligible to receive a rating from the Energy Star program, at least 50 percent of a building's floor area must be defined by one of the eligible space types, which assigns the building to a peer group against which the facility will be compared. Based on their space type, geographical location, and level of business activity, the program assigns each facility a national energy performance rating on a scale of 1 to 100. Facilities that meet certain criteria and achieve a rating of 75 or better are eligible to apply for an Energy Star designation.

Studies by the EPA show that the more than 3200 buildings nationwide that have earned the Energy Star rating use about 35 percent less energy than comparable buildings.

National Institute of Building Sciences (NIBS) Whole Building Design Guide

The *Whole Building Design Guide* is a collection of sustainable building resources openly available on the Internet at: www.wbdg.org. An entire section is devoted to construction and demolition waste management (www.wbdg.org/resources/cwmgmt.php), including articles and links to a variety of other resources, most notable of which is a database of C&D-related services.

Created in 2002 with funding from the U.S. General Services Administration, the Construction Waste Management Database contains a list of C&D waste recyclers and haulers, grouped by zip code. The database can be accessed through the NIBS Design Guide Web site at www.wbdg.org/tools/cwm.php.

Green Building Initiative Green Globes® Program

The Green Building Initiative (GBI) is a consortium of industry, government, and nonprofit representatives who have successively modified an early Canadian program into an online resource that is promoted as a more streamlined and interactive alternative to LEED. Accredited as a national standards developer by the American National Standards Institute (ANSI) in 2005, the Green Globes program uses a standard template for all new construction and major renovations, regardless of size or building type.[4] The system is questionnaire-based, and provides early feedback through electronic reports, including recommendations during schematic design and construction documents to aid users in improving their projects.

Feedback provided from the questionnaires includes:

- An initial Green Globes rating for the project
- A recap of building highlights
- Recommendations for sustainable improvements
- Supplemental information and links for use by the designer/builder

The online program assesses sustainability in seven projects areas: (1) site, (2) water, (3) resources, (4) emissions, (5) project management, (6) energy, and (7) indoor environment. Points are awarded for meeting criteria in the various areas comprising each category. Here is a general breakdown of the criteria:

1 *Project management (50 points)*: Integrated design process, environmental purchasing, commissioning (plans for systems testing after construction), emergency response plan
2 *Site (115 points)*: Development area, ecological effects (erosion, heat island, light pollution), watershed features, site ecology enhancement
3 *Energy (360 points)*: Energy performance, reduced demand (space optimization, microclimatic design, daylighting, envelope design, metering), energy efficiency features (lighting, heating, and cooling equipment), renewable energy (solar, wind, biomass), transportation
4 *Water (100 points)*: Water performance, water conserving features (equipment, meters, irrigation systems), on-site treatment (stormwater, greywater, blackwater)
5 *Resources (100 points)*: Low-impact systems and materials (LCA), minimal use of nonrenewables, reuse of existing buildings; durability, adaptability, and disassembly; demolition waste (reduce, reuse, recycle), recycling and composting facilities
6 *Emissions, effluents, and other effects (75 points)*: Air emissions (boilers), ozone depletion, sewer and waterway protection, pollution control (procedures, compliance with standards)
7 *Indoor environment (200 points)*: Ventilation system, indoor pollution control; lighting (daylighting and electric), thermal comfort, acoustic comfort

The program addresses C&D waste in the *Resources* category, where it awards points for the reuse of existing buildings and diversion of construction site waste from landfills.

Green Globes incorporates product Energy Star ratings and life-cycle assessment information into their program, and uses GBI's ANSI process for updating technical criteria.

Projects are awarded one to four globes, depending on the number of criteria met through the questionnaire.

Green Globes requires third-party assessment of the construction documents and a walk-through review of the project at final completion. Assessors are certified under a separate program run by CSA America, Inc.

Green Globes certification also takes a long-term approach, providing a module for analysis of building performance data after 12 months, including recommendations for further operational improvement. The program encourages five to eight asset management strategies, and provides for recertification of existing buildings after three years of operation. Costs for the program include GBI membership costs (ranging from $500 for a single building to $2500 for unlimited buildings) and the cost of third-party verification for each project ($4000 to $6000). GBI does allow self-assessment by the builder, but this option does not permit the project to be identified as a Green Globes project. Information on this rating system is available at the Green Building Initiative Web site at www.thegbi.org, or by contacting GBI at 877.GBI.GBI1.

Florida Green Home and Green Commercial Buildings Standards

FLORIDA GREEN HOME STANDARD

The green home standards define basic criteria by which a Florida home, new or existing, can be designated green under their program. Certifying agents can guide designers, builders, or homebuyers through the process of qualifying and documenting green homes. Florida green home designation is administered by the Florida Green Building Coalition, Inc. (FGBC) (www.floridagreenbuilding.org). FGBC's stated goal is to provide designers and contractors with the tools and guidance necessary to help them design more energy-efficient homes and commercial buildings.

FLORIDA GREEN COMMERCIAL BUILDING STANDARD

The intent of FGBC's commercial standard is to encourage owners of smaller commercial projects to adopt green and sustainable strategies during the design and construction of their project and to receive recognition for their efforts. The commercial building designation standard covers all commercial occupancies listed in the current Florida

Figure 8.2   **States operate a wide variety of green and energy-efficient residential certification programs.** © 2010, Mark Rasmussen, BigStockPhoto.com.

Building Code. To use this standard, the project design team must review the FGBC checklist and reference guide to determine which program credit points to pursue. The owner of the project must designate one of the design team members to be the designated professional, who is responsible for compiling the appropriate documentation for each credit point the design team is pursuing under the guidelines. See Fig. 8.2.

GreenPoint Rated New Home Program (California)

The *GreenPoint Rated* New Home Program is a program developed by Build it Green, a Berkeley, California nonprofit association. The program requires third-party verification that new home construction has met the program requirements for sustainability and energy efficiency in the areas of indoor air quality, energy and water efficiency, and resource conservation.

If the new home meets minimum point requirements in each category, and scores at least 50 points on either the Single Family or Multifamily GreenPoint Rated Checklist (verified by a Certified GreenPoint Rater), the home will earn a GreenPoint Rated label. The GreenPoint Rated label provides a numerical score that allows homebuyers to compare the environmental performance of different homes. The GreenPoint program also offers a link for local governments to incorporate green building requirements into their local ordinances. The program is closely linked to the California energy and building code requirements. More information is available on the Build it Green Web site at www.builditgreen.org.

U.S. State Green Building and Energy-Efficiency Programs

Alabama

EarthCraft House: www.earthcrafthouse.org/

Energy Right: www.energyright.com

Arizona

Scottsdale Green Building Program: www.scottsdaleaz.gov/greenbuilding

TEP Guarantee Home: www.tucsonelectric.com/Green/GuaranteeHome/ aboutguarantee.asp

California

Build it Green (see detailed description in this chapter): www.builditgreen.org

California Green Builder: www.cagreenbuilder.org/

Earth Advantage: www.earthadvantage.com/

Santa Monica Green Building Program: www.smgov.net/departments/ose/categories/ buildGreen.aspx

San Jose Green Building Program: www.sanjoseca.gov/esd/natural-energy-resources/ gb-background.htm

Colorado

Built Green Colorado: www.builtgreen.org/

City of Boulder Green Points Program: www.bouldercolorado.gov/index.php

Florida

EarthCraft House: www.earthcrafthouse.org/

Florida Green Building Coalition (See detailed description in this chapter): www. floridagreenbuilding.org

Good Cents: www.goodcents.com

Georgia

EarthCraft House: www.earthcrafthouse.org/

Energy Right: www.energyright.com

Good Cents: www.goodcents.com

Right Choice: www.jacksonemc.com/RightChoice.rightchoice.0.html

Hawaii

Hawaii BuiltGreen: http://biahawaii.org/

Kentucky

Energy Right: www.energyright.com

Louisiana

Power Miser Homes: www.cleco.com/site461.php

Michigan

Green Built Michigan: www.greenbuiltmichigan.org/

Minnesota

Triple E New Construction: www.mnpower.com/powerofone/

Mississippi

Energy Right: www.energyright.com

New Jersey

New Jersey ENERGY STAR Homes: www.njcleanenergy.com/residential/programs/programs

North Carolina

Energy Right: www.energyright.com

NC HealthyBuilt Homes: http://healthybuilthomes.org/

Oregon

Earth Advantage: www.earthadvantage.com/

South Carolina

EarthCraft House: www.earthcrafthouse.org/

Good Cents: www.goodcents.com

Tennessee

EarthCraft House: www.earthcrafthouse.org/

EcoBuild: www.mlgw.com/SubView.php?key=about_ecobuild

Energy Right: www.energyright.com

Texas

Austin Green Building Program: http://www.austinenergy.com/Energy%20Efficiency/Programs/Green%20Building/index.htm

Build San Antonio Green: www.buildsagreen.org/

Frisco Green Building Program: www.ci.frisco.tx.us/departments/planningDevelopment/greenbuilding/Pages/default.aspx

Good Cents: www.goodcents.com

Vermont

Vermont Builds Greener: www.veic.org/ProjectProfiles/VermontBuiltGreen.cfm

Virginia

Arlington County Green Home Choice Program: www.arlingtonva.us/Departments/EnvironmentalServices/epo/EnvironmentalServicesEpoGreenHomeChoice.aspx

EarthCraft House: www.earthcrafthouse.org/

Washington

Built Green Washington: www.builtgreenwashington.org/

Built Smart: www.ci.seattle.wa.us/light/conserve/resident/cv5_bs.htm

Earth Advantage: www.earthadvantage.com/

Wisconsin

Green Built Home: http://wi-ei.org/greenbuilt/

International Programs

Two large international programs merit mention for architects and contractors dealing with international projects.

UNITED KINGDOM: BREEAM

The BREEAM program in the United Kingdom has certified more than 100,000 buildings to date, with another 500,000 registered in the program. Certification under this program requires a building to obtain performance credits across nine categories, for an overall score of pass, good, very good, excellent, or outstanding. The Gulf States BREEAM rates buildings with one to five stars.

BREEAM certification for new facilities typically costs no more than $6000. BREEAM's *In-Use* assessment for existing buildings can cost between $1000 and $34,000, depending on the number of assets assessed. Compiling the information for the program's assessment report should normally take 4 to 5 days, while certification can take as little as 10 days.

For international certification, BREEAM is working with local Green Building Councils (in the Gulf States and in Europe) and has developed BREEAM Bespoke and BREEAM International, which are separate from the United Kingdom rating organizations.

AUSTRALIA: GREEN STAR

Modeled on BREEAM, Australia's Green Building Council's Green Star rating has been awarded to more than 140 buildings, with another 500 in the process of being certified.

To obtain a Green Star Rating, a building must achieve a minimum score across nine categories. A percentage score is used to calculate the categories, and the program uses weighting factors to assess the overall Green Star rating. The environmental weighting factors vary across states and territories to reflect differing environmental concerns in various regions of Australia.

While an owner or design professional can use program tools to self-assess his project, to claim certification and qualify for the Green Star label, each project must be verified by the Green Building Council of Australia. Certification fees can range between $4000 and $20,000. Once rather complex to follow, a recent overhaul of the program guidelines streamlined the requirements and reduced the overall program fees by approximately 15 percent. The certification process requires two rounds of assessment, which together can take 6 to 18 months to complete.

References

1. "ICC Evaluation Service, Inc. (ICC-ES)." International Code Council Evaluation Service (ICC-ES). January 3, 2010 <http://www.icc-es.org/>.
2. "International Green Construction Code." International Code Council (ICC). January 3, 2010 <http://www.iccsafe.org/.>.
3. "NAHB National Green Building Program." National Association of Home Builders. November 20, 2009 <http://www.nahbgreen.org/>.
4. "Green Globes." Green Building Initiative. November 28, 2009 <http://www.greenglobes.com/>.

9

DOCUMENTING COMPLIANCE

Maintaining and collecting documentary evidence of the amount of waste sent to land-fills and diverted to recycling is essential for the contractor to establish compliance with his waste management plan. As noted earlier in this book, the recycling rate for the C&D waste on a construction project is often set by municipal ordinance or via contract requirements with the project owner. It is therefore a legal and contractual requirement that the contractor be able to document that he has met the requirements. A failure to do so can cause severe legal problems for him and his owner, and in extreme cases delay the certificate of occupancy for the project.

There are other reasons for the contractor to be concerned with appropriate waste management documentation. To measure his own performance and profitability, the contractor should want to keep accurate data on the costs of recycling waste. These values represent competitive intelligence. They are no different, and no less valuable, than accurate labor, material, and equipment rental costs. The more accurately a contractor can estimate his waste management costs, the more tightly he can bid on future projects and the more likely he is to win those bids.

As recycling becomes more commonplace in construction projects, software and online resources will be developed to assist the contractor in documenting the jobsite recycling rate. In fact, WasteCap Resource Solutions (www.wastecap.org), a nonprofit recycling advocacy group, has already unveiled an onsite recycling documentation to aid contractors in tracking and documenting the reuse and recycling results on construction, deconstruction, and remodeling projects. Called *WasteCapTRACE*™, the new program claims to help streamline the documentation process for contractors.

The contractor's project manager should have four main goals in mind as he sets up the documentation requirements for his waste management plan:

1 *Define the measure*: Remind all subcontractors and markets to report by weight in pounds—not by volume or tonnage. Weight measured in pounds provides a higher degree of accuracy than weight in tons, and may make the difference on projects with high required recycling rates. The contractor can always convert pounds to tons on his reporting forms for simplicity of reporting.

2 *Who will measure?*: Outside verification is always preferable to on-site or contractor-controlled scales or other means. Municipalities and owners increasingly want independent confirmation of a contractor's waste management compliance.

3 *Who will document?*: Which person is responsible for documenting compliance from each firm, and how often must he do so? (Monthly is recommended.)

4 *How to document?*: See this book's Online Resources for a collection of documentation forms. The form is not as important as the clarity of the information, the date, and the attachments supporting the information.

Large contractors with office personnel to support the management of a recycling plan have an advantage in managing the necessary documentation. Even these contractors, however, will struggle to track documentation that is not organized ahead of time. For small contractors unable to commit office personnel to assist the field superintendent and project manager in documenting the recycling program, creating a clear and organized program of documentation tied to monthly applications for payment is essential.

Contractor-Generated Information

The management of C&D waste on a site requires the general contractor to document the progress of the plan against the goals laid out in the waste management plan or compliance documents he filed with the owner and architect. This requires two types of documentation:

1 Interim or periodic reports (normally provided with each periodic application for payment)

2 Final report (provided as part of the closeout documents)

The interim report contains waste management activity up to a certain cut-off point each month. This cutoff point should coincide with the billing cutoff day to simplify recordkeeping and tracking. As with any other information required of the subcontractor, providing waste management documentation to the contractor each month should be a requirement of his contract, with the provision that the contractor's monthly payment to the subcontractor will be held until it is provided.

Interim reporting normally consists of a simple update to the waste management plan, showing the total waste created, waste sent to a landfill or incinerator, and waste diverted (recycled or reused). A periodic recycling or diversion rate is also included to demonstrate whether the contractor is on track with his overall goals (see this book's Online Resources for a sample periodic form).

Here is an example of the information provided on an interim report:

- *Waste management plan diversion goal*: 50 percent of total waste
- *Total estimated C&D waste:* 74.5 tons (67.6 metric tonnes)
- *Total C&D waste generated to date*: 31.35 tons (28.44 metric tonnes) (46.4 percent of total estimated)

- *Waste by category*
  - *Asphalt*: 7.7 tons (7.0 metric tonnes) generated/4.7 tons (4.3 metric tonnes) recycled/ 3.0 tons (2.7 metric tonnes) reused
  - *Concrete*: 9.4 tons (8.5 metric tonnes) generated/7.5 tons (6.8 metric tonnes) recycled/ 1.2 tons (1.1 metric tonnes) reused/0.3 tons (0.27 metric tonnes) landfilled
  - *Metals:* 2.6 tons (2.4 metric tonnes) generated/2.6 tons (2.4 metric tonnes) recycled
  - *Gypsum wallboard*: 2.4 tons (2.2 metric tonnes) generated/0.4 tons (0.36 metric tonnes) recycled/2.0 tons (1.8 metric tonnes) landfilled
  - *Insulation*: 0.2 tons (0.18 metric tonnes) generated/0.08 tons (0.07 metric tonnes) recycled/0.12 tons (0.11 metric tonnes) landfilled
  - *Wood*: 1.2 tons (1.09 metric tonnes) generated/1.0 tons (0.9 metric tonnes) recycled/ 0.2 tons (0.18 metric tonnes) reused
  - *Masonry*: 5.9 tons (5.4 metric tonnes) generated/5.9 tons (5.4 metric tonnes) recycled
  - *Mixed waste*: 1.5 tons (1.36 metric tonnes) generated/0.9 tons (0.82 metric tonnes) recycled
  - *Architectural salvage*: 0.4 tons (0.36 metric tonnes) generated/0.4 tons (0.36 metric tonnes) reused
  - *Packaging reduction*: 0.05 tons (0.04 metric tonnes) would have been generated/ 0.05 tons (0.04 metric tonnes) reduced
- *Total C&D waste reduced at source*: 0.05 tons (0.04 metric tonnes)
- *Total C&D waste reused*: 4.8 tons (4.35 metric tonnes)
- *Total C&D waste recycled*: 23.08 tons (20.94 metric tonnes)
- *Total C&D waste diverted*: 27.93 (25.34 metric tonnes)
- *Total C&D waste sent to landfill*: 3.42 tons (3.10 metric tonnes)
- *Diversion rate*: 73.6 percent

Depending on the preference of the architect and owner, interim reports may not require backup documentation. The lag in obtaining backup information from various sources can make it difficult to fully document an interim report. A contractor, for instance, may be aware of the weight of container waste through electronic or verbal reports, but not receive the paper documentation until sometime later.

Final documentation submitted at the end of the project must, of course, contain full documentation to support the recycling or diversion rate claimed by the contractor. The quality and detail of documentation required by the auditor, whether architect or municipality, will vary. Generally, however, the reviewers of construction waste documentation realize that contractors are dealing with large volumes and weights of material in a fragmented industry, and that a wide range of documentation quality is inherent in the industry. There is, or should be, some leeway in making good-faith estimates of the weight of materials where no documentation can be had. Following are some general tips for field superintendents regarding field documentation they can manage to support their firm's waste management plan:

1 *Photograph as much as possible*: When time permits, photograph (with date stamps) fully loaded containers. In the event weight slips disappear or never show up, the contractor at least has a photograph of the volume of material, leaving only the weight to argue over.

2 *Log pickups and container empties in the daybook*: This is routine for field superintendents, and is invaluable in tracking missing weight slips from recyclers.

3 *Keep paper forms on hand*: Record jobsite pickups by nonprofits or direct purchasers. Use the form to fully describe the item and its size and weight (if known). Have the purchaser sign the form.

The final report submitted by the contractor to the architect and owner must contain all the necessary backup paperwork required to document the recycling rate claimed by the contractor. Here is a list of types of generic backup information that are normally acceptable to reviewers:

- Date-stamped photographs of recycled material volume (scale elements in the photograph are helpful)
- Weight tickets from a certified or public scale
- Weight tickets from recycler or end user
- Hauler tickets
- Field superintendent daily log
- Independent witnessed field scale or crane scale weights (photographs are beneficial as well)
- Signed recipient statements of donated goods
- End user verifications of any type
- Letter from distributor or manufacturer documenting packaging reduction
- Architect, engineer, or owner representative observation of volume or weight of recycled or stored materials (note in field observation report or on contractor form)

INFORMATION FROM SUBCONTRACTORS

Demolition Subcontractors Demolition subcontractors are largely adept at weighing truckloads of materials and producing documentation that satisfies the contractor's waste management plan needs. The main concern of the general contractor is ensuring that the demolition contractor's materials documentation lines up cleanly with his waste management plan so interim and final reporting can be accomplished with a minimum of translation time. Since the types of recycled materials can change during the course of construction, this is an argument for the contractor to set up his initial waste management plan with broader categories that allow him some flexibility later on.

Demolition subcontractors may also be able to fully manage and track the contractor's waste management compliance for him as an added service.

Trade Subcontractors Some subcontractors, including plumbing, electrical, and roofing subcontractors, may have well-established construction waste recycling efforts that they include in every contract they perform. Such an initiative by a subcontractor can be very helpful to a contractor in that it removes the burden of dealing with that particular line of waste from his overall C&D waste responsibilities. This is all for the best, as long as the subcontractors are willing and able to provide sufficient documentation to the general contractor to enable him to count this waste toward his overall recycling rate.

Architectural Salvagers Architectural salvagers deal mostly with two numbers: wholesale value and retail value. In projects with high-value salvage, the contractor may net some actual income from the salvager. In other projects with less valuable salvage, the best he may do is obtain the removal of some of his demolition at no cost by the salvager. This will save some money on the overall demolition cost, but probably not a significant sum (see Fig. 9.1).

Tip Box

- Bath Fixtures
- Brick and Ornamental Masonry
- Doors and Hardware
- Equipment and Machinery
- Flagstone
- Floor Registers
- Flooring
- Furnishings
- Glazing (Stained, Bullet, or Float)
- Ironwork or Ornamental Metalwork
- Hardware
- Lighting
- Lighting Fixtures
- Millwork, Including Standing and Running Trim
- Mirrors
- Ornamental Woodwork
- Mantels
- Paneling
- Pavers
- Roof slate
- Signs and Advertising
- Stair Parts
- Windows
- Wood Timbers

Figure 9.1 Examples of architectural salvage.

None of this is particularly helpful to the contractor attempting to document his recycling rate by weight or volume. He needs the architectural salvage company to provide documentation to him as a listing, either in bulk or by item. If the contractor has arranged for his subcontractors to use an independent certified scale company, he can simply ask the salvager to provide documentation from this company of the weight of salvaged materials on his truck. Otherwise, the salvager needs to provide the contractor with an itemized statement of the salvage items and their individual weight. See this book's Online Resources for a sample statement.

Outside Information Management

Contractors will want to treat recycling reporting like any other form of subcontractor-required documentation. On projects with monthly pay applications, the recycling report from a subcontractor should be just another part of the package of paperwork he submits in support of his application. Just as contractors require certified payroll, partial release of liens, or other documents with each pay application, they should require a monthly summary of recycling activities. Because most of the recycling activity on the jobsite will be run through the contractor-controlled recycling center, this requirement will typically only affect a few subcontractors. The demolition contractor may have independent reporting that he must do for any recycling operations directly under his contract. Similarly, electrical, plumbing, or masonry subcontractors may handle recycling of waste associated with their trades directly under their contracts. They would need to report this activity to the general contractor for inclusion in the overall compliance report.

Where this situation becomes difficult is with suppliers who voluntarily reduce packaging or product-related waste. Measured in weight, this reduction may not be a considerable part of the overall recycling rate as it tends to be lightweight cardboard, foam cushioning, and plastic wrapping. Nevertheless, the reduction represents a true diversion as long as the reduced packaging does not make its way to a landfill through other means, and enough of these types of diversions from landfills may be the difference-maker on certain projects. As an added complication, the supplier holds a contract with the subcontractor, not with the general contractor, and is therefore less likely to respond to a general's demand for information.

The best solution in these cases is to seek the cooperation of the subcontractor in obtaining basic packaging reduction documentation from the supplier (see this book's Online Resources for a sample document).

Collecting information from other parties can be more bothersome. Nonprofit organizations are notorious for their unresponsiveness and poor document management. Contractors are best advised to prepare their own documentation, listing the types and weights of donated materials, and have a representative of the nonprofit organizations sign and date the form. If the contractor himself is hauling the material to the nonprofit organizations' location, as is often the case, he can arrange for independent weight verification of the donation. Otherwise, the list of materials and a photograph should serve as sufficient evidence of the donation, and a reasonable weight estimate of furnishings or framing can be made (as long such an estimate is acceptable to the owner or municipality).

Established recyclers are reliable record-keepers, and know the importance of providing timely weight and diversion records to the contractor. The only difficulty in most contractor/recycler relationships is in the area of mixed loads, where several complications can occur:

- Mixed loads may sit in the yard for a prolonged period before they are sorted.
- A particular contractor's load will likely be mixed with many other project loads before sorting.
- Regardless, it is not possible in most facilities to accurately measure the recycling rate of a particular load.
- Mixed recyclers have different personnel and equipment, so diversion rates can vary wildly for mixed loads—from 10 percent to 90 percent.
- A mixed recycler should be able to tell a contractor his average diversion rate over time, but he can rarely tell him what his particular load's diversion rate will be.
- A contractor who feels he has high-quality mixed loads, or ones that should yield a high recycling rate, should shop carefully for a recycler with the equipment to take full advantage of the loads.

WEIGHT DOCUMENTATION

Trucking company scales or on-site weighing systems may or may not be accepted by a public body as reliable documentation of the weight of materials. On a large project where the owner has full-time site representation (clerk-of-the-works or otherwise), on-site scales may be accepted when the owner's representative can witness the weight measurement. Otherwise, the contractor and owner should agree to a reasonable standard for documenting weight. Since the values in question are not net payments, but a reduction of net costs, the owner and municipality may well be satisfied with an outside weight verification from the recycler, hauler, or anyone else not directly associated with the contractor. Otherwise, the hauler needs to obtain a *tare or unladen weight*, a certified statement of the weight of an empty vehicle.

Most commercial haulers will possess vehicle weight certifications (vehicle excise or design weight certification), documenting the empty weight of the truck. The contractor will also need some documentation—or reasonable estimate—of the weight of the container.

Public scales certified by state departments of motor vehicles are widely recognized, and considered reliable. They provide certified weight certificates, including the vehicle's gross weight, vehicle identification number, and type of vehicle. Most weigh houses are open 6 days per week all year long. For busy areas, an appointment is recommended, but contractors may establish an open contract with the weigh station to accept vehicles during certain hours. Scales can often be found at area truck stops, though the quality of documentation varies widely since these scales are mostly intended as weight checks for truckers. A truck stop scale locater is available on *www.bigrigjobs.com/Truck-Scales* and on *www.interstatescales.com*, two commercial Web sites and 1200 nationwide CAT scale locations can be found on that company's Web site at *www.catscale.com*. In California, check *http://publicscales.org/* for the location of certified scales in the project area.

Crane Scales   Crane scales can be very useful in documenting on-site wastes of large pieces of equipment removed from roofs or otherwise lifted by a crane. Crane scales are essentially strain gauges, which measure the overall weight of the load being lifted by the crane. If the crane is lifting material in a container or on steel plate that is not being recycled, the contractor needs to subtract out that weight from the amount claimed for recycling or reuse. Documentation for on-site crane weights can be difficult. Photographic documentation of the overall lift of a large item may not be able to clearly show the weight on the crane scale. A separate photograph showing the strain gauge reading should be acceptable. Another option would be a combination of a photograph and independent certification by the architect or other owner representative, which would be ideal, but this is not always possible.

Special Items Documentation

Some items cannot be easily weighed, and others are purchased and never pass over a scale. Wood and concrete chipped or milled on site may be sold directly to markets and are never weighed by the purchaser. Equipment sold directly from the jobsite may be picked up by the buyer and directly transported to his facility.

As previously noted, it is a wise move for the contractor to arrange an open contract for his project with an independent certified scale, enabling any recycler to visit the scale and obtain a weight report on his behalf. The contractor can then insist on a weight report as a condition of the contract, and give a purchaser or market little excuse not to provide it.

Even with such a service, however, there will be times when recycled material escapes the jobsite with no recorded weight and no reliable way of obtaining it. In these instances, it is very helpful to have a photograph of the recycled material on the ground or in a container. Material weights of recycling can be estimated reasonably enough if a photograph exists to document the existence of the material and its general nature. Failing this, the contractor should demand from the recycler a written statement of type and approximate volume of the material. If he wants to estimate the weight as well, this should be encouraged. At the very least, however, the contractor should seek to obtain as much information on the purchaser's letterhead as possible as this represents an outside party with no particular interest in aiding the contractor in meeting his recycling rate requirements.

MATERIAL REUSED ON-SITE

Material that is processed or reused on-site, such as crushed concrete and brick waste or wood waste, is not normally weighed prior to reuse unless the contractor has rented a site scale. The weight of this material can be reasonably estimated by container volume or stockpile size, however. Since volume and weight of materials are subject to dispute by a reviewer, the field superintendent should document the volume of material as well as he can to remove at least this one factor from dispute. A photograph with a scale element, or independent measure by the architect or civil engineer on a field observation visit, should be adequate to verify the approximate volume of material. Then it is only a matter of agreeing to the weight of the material per cubic yard.

RESOLD OR DONATED ITEMS

Equipment sold directly from the jobsite to an end user should be documented by independent scale weight whenever possible. If the equipment consists of known manufacturer's model numbers, or large equipment that is identifiable in the industry, its weight may be reliably estimated from published tables or manufacturer's information. If handled by a crane, use of a crane gauge to document the weight is also a possibility.

10

WASTE MANAGEMENT IN THE
CONSTRUCTION DOCUMENTS

Architects and other design professionals face a dilemma in addressing recycling and solid waste management issues in the construction documents. Where the owner has stipulated (or the local municipality has mandated) a certain recycling rate, the architect should include this requirement in the specifications. But because the way in which the contractor achieves this goal is part of the so-called *means and methods* of construction, the design professional needs to steer clear of defining in any detail how he must accomplish the requirement. This is true for professional liability purposes, but also because the contractor is best qualified to assess the nature of the waste generated on the project and how to market it to recyclers.

Architects and other design professionals can do a great deal within the construction documents to boost the recycling rate on a project. Following are some simple tips for architects:

See Fig. 10.1 for tips on how design professionals can minimize construction waste generation in the construction documents.

1 *Economize materials*: Perform multiple functions with one material rather than using a number of products to meet a single need. Design HVAC and plumbing systems to optimize the number of functions they can perform.

2 *Use efficient areas and volumes*: Avoid creating small spaces and enclosures that do not serve the client's program needs. Whenever less material is required by the design, less waste is generated at the jobsite.

3 *Use standard material and product dimensions*: Design with standard framing and board sizes in mind to minimize scrap waste or cut-offs, at the jobsite.

4 *Limit field shoring and bracing*: Where possible, select construction systems that do not require temporary support, shoring, construction aids, or other materials that will be disposed of as debris after their use in the project.

5 *Limit adhesives*: Avoid using materials that rely on adhesives. The extensive use of adhesives on a project generates a great deal of residue and container waste. Adhesives also make end-of-life building recycling more difficult.

- Economize materials.
  - Perform multiple functions with one material.
- Use efficient areas and volumes.
  - Combine spaces; reduce uncommitted space.
- Use standard product dimensions and lengths.
  - Design around modular sizes.
- Limit field shoring and bracing.
- Limit the use of adhesives.
- Use products with integral finishes.
- Specify products that do not deteriorate before installation.

Figure 10.1   Minimizing waste in the construction documents.

6 *Use integral finishes*: Select materials with integral finishes. Limit requirements for applied finishes, laminates, and coatings, all of which generate more jobsite waste.

7 *Specify jobsite durable products*: Avoid specifying materials that are particularly sensitive to site damage, contamination, or exposure, all of which increase jobsite waste.

Both the practical and financial considerations of identifying markets for C&D waste make it necessary for the contractor to have enough freedom to negotiate with haulers and modify the recycled products.

For this reason, architects need to limit the language in the construction documents to reflect the following goals:

- State the required recycling rate to be achieved (if the owner or municipality has mandated a rate).
- Include a specification section requiring an initial waste management plan, periodic (usually monthly) progress reports, and a final report documenting the recycling rate.
- Include basic provisions in the drawings and specifications that will enable the contractor to meet his goals. Specifically, this refers to anticipation in the civil documents that a recycling zone will be created and how it might be configured.

Specifications

The specifications prepared as part of the construction documents provide the architect with the opportunity to define the overall scope and nature of the recycling effort on the site. In preparing the specifications she will attempt, as she does in other technical specification sections, to stake out the overall goals and requirements of the program without telling the contractor exactly *how* to manage the program. The *how*, of course, are the means and methods of construction, which are the sole province of the contractor. The closest an architect will get to intruding on this turf is to define in the specifications the rough outline of the waste management plan. She includes this section in the specifications to ensure that the contractor includes the bare minimum of key points in the plan he prepares for his own use in actually managing the recycling program on his construction site.

Specification sections that address aspects of waste management or recycling:

- *Section 00120*: Supplementary Instructions to Bidders—Resource Efficiency
- *Section 00800*: Supplementary General Conditions
- *Section 01010*: Summary of the Work
- *Section 01030*: Alternates
- *Section 01031*: Waste Management and Recycling Alternates
- *Section 01060*: Regulatory Requirements
- *Section 01094*: Definitions
- *Section 01200*: Project Meetings
- *Section 01300*: Submittals
- *Section 01400*: Quality Control
- *Section 01500*: Construction Facilities and Temporary Controls
- *Section 01505*: Construction Waste Management
- *Section 01600*: Materials and Equipment
- *Section 01630*: Substitutions
- *Section 01700*: Contract Closeout

Section 01505 is the specification section that deals most closely with waste management issues. This section will contain the most detailed requirements the contractor must follow in managing C&D waste from the project. Following are the general components of this construction and demolition waste specification section:

- Intent of the waste management goals for the project
- Draft waste management plan
- Materials for which recycling is required
- Final waste management plan
- Waste management plan implementation
- *Reporting requirements*: interim and final

See this book's Online Resources for sample C&D waste specification sections.

The architect's specifications will typically state the required components of the final waste management plan. Most often, these minimum requirements are:

1 *Review*: Analysis of the proposed jobsite waste to be generated, including types and quantities.
2 *Landfill options*: The name(s) of the landfill(s) where trash will be disposed of, the applicable landfill tipping fee(s), and the projected cost of disposing of all project waste in the landfill(s).
3 *Alternatives to landfilling*: A list of the waste materials from the project that will be separated for reuse, salvage, or recycling.
4 *Meetings*: A description of the regular meetings to be held to address waste management.
5 *Materials handling procedures*: A description of the means by which any waste materials identified in item 3 above will be protected from contamination, and a description of the means to be employed in recycling the above materials consistent with requirements for acceptance by designated facilities.
6 *Transportation*: A description of the means of transportation of the recyclable materials (whether materials will be separated on-site and self-hauled to designated centers, or whether mixed materials will be collected by a waste hauler and removed from the site) and destination of materials.

CONTRACTORS AND SPECIFICATION CONFLICTS

Architects are professional and well intentioned. When they write specifications, they are genuinely interested in providing the contractor with clear guidance regarding the intent and requirements of the recycling program. The specification language, of necessity, needs to be precise and strict to clearly convey the owner's intent.

Sometimes, however, this need for precision comes into conflict with the realities of how C&D waste is actually managed on a construction project. These types of conflicts typically occur in several areas:

■ *Extent of recycled products*: Conflicts occur in cases in which the construction specifications require the recycling of a product for which there is no market.
■ *Specific container and management requirements*: Conflicts occur when the design professional strays into the means and methods of construction by stating specific container sizes and requirements for the recycling zone.
■ *Bid requirements*: Architects occasionally require bidding contractors to submit waste management plans as part of their bid, and to propose a recycling rate that they can meet. This is practical in new construction, but impractical in demolition and renovation work, where a complete site assessment is necessary to estimate the recycling rate.

Another type of problem that occasionally occurs is when owners realize that salvage items in their facility have more value than they knew. The bid documents may not be explicit in stating whether the owner or contractor is entitled to the value of architectural salvage, and owners may claim that value for themselves once they realize that architectural salvage is one of the few areas where recycling money may actually flow into the project coffers. When this issue occurs, there is no easy solution other than for both parties to clearly state their position and either negotiate or mediate a resolution. Contractors who have experienced this issue previously, or see the potential for such a claim occurring, should seek to clarify the issue in pre-bid requests for information or in specific language in the owner/contractor agreement.

The best method of handling these types of conflicts is for the contractor to ensure that the architect is an integral part of the recycling team from the very beginning. If an owner's representative is available, it is advantageous for that person to be included as well. When these individuals are present at the initial on-site assessment, for instance, they are able to see for themselves the difficulties affecting compliance with the bid or project requirements, and are much more likely to agree to measures that adjust those requirements to better meet the circumstances.

Drawing Information

The construction drawings prepared by the architect can aid the demolition contractor by clearly stipulating the building components to remain. In the case of structural elements, this may be abundantly clear in the structural and architectural drawings. In the case of secondary elements, such as doors, windows, finishes, and fixtures, the extent of demolition may not be at all clear to the demolition and general contractors unless the architect clearly designates it. Architects need to be sensitive to the contractor's problem of needing to identify the major and minor parts of the work for demolition and recycling bids. In particular, architects should ensure that the following is clear in their documents:

- *Site plans*: Indicate existing trees and shrubs to be retained or removed. Show extent of grading, seeding, or sodding.
- *Structural plan*: Indicate foundation elements that are to be retained or supplemented. Clearly designate existing steel or masonry to remain versus new sections.
- *Selective demolition plan and elevations*: It is always helpful to prepare a specific plan and elevation to designate the areas of renovation project to be removed. These drawings help the architect to better cover the new scope in his documents, and are invaluable to the contractor in quickly assessing the scope of demolition.
- *Architectural slab or foundation plan*: Where architects are aware of the specific needs for cutouts in existing slabs, or wish to limit the extent of cutouts, prepare a slab plan to communicate this intent.
- *Floor plan*: Clearly indicate new walls versus existing to remain with shading or different linework.
- *Finish plan or schedule*: Designate types of finishes to be protected and reused. Where only portions of finishes are to be retained, create a finish plan indicating where demolition occurs.

■ *Exterior elevations*: Indicate the exterior finishes to remain. If new construction abuts existing, indicate what happens to the exterior finishes enclosed within the addition.

Contractors often face the difficulty of determining which specific pieces of the existing construction are required to remain and which are subject to demolition and recycling. This may not be a huge cost issue during bidding, but becomes critical as selective demolition work begins in a renovation project. Involving the architect in the recycling assessment meeting, or requesting a predemolition walk-through, can alleviate much of this dilemma.

Site Plan Information

The location and configuration of the recycling zone on a project is the responsibility of the general contractor. This fact is well established. However, the construction documents include site planning information that affect (or limit) the contractor's ability to perform this work. In most areas, the civil engineer is required to file a soil erosion and sedimentation control drawing, which may be submitted for approval during the bidding phase of the project, or prior to the selection of the contractor. If this plan does not anticipate the creation of a recycling zone, does not include a sufficient area to accommodate the project needs, or does not include the recycling zone in the protected area, then it will need to be refiled. Other complications occur when the contractor wants to temporarily grade a portion of the site for relatively level recycling storage and access. Even though construction activities result in any number of grading adjustments during the course of the project, the regrading of a large area for a period of several months may be a situation that a soil erosion officer will want reflected in the plan as a construction phase. Changes in the soil erosion and sedimentation control plan may delay the start of the project.

All of these situations cannot be known by the civil engineer early in the project cycle. They can be anticipated, however, so that some provision for the waste management needs of the project are reflected in the civil documents. Even if changes to the approved plans are required, the extent of such changes is limited. The reviewing official is not being asked to evaluate a new feature of the plan—only a modification to a feature he has already approved.

Recycling zones also require stabilized aggregate or stone bases for truck traffic, a situation that must normally be addressed in the site planning with the extent and drive detail indicated.

Site plans prepared by a civil engineer normally show one or more construction entrances to a site. Because of the size of recycling roll-off containers, a looped entrance and exit may be required on tight sites to enable efficient drop-off and retrieval of roll-off containers. Such dual entrances and exits, placed closely together on a heavily traveled road, may be a source of some concern to local officials. Although temporary construction traffic patterns may be granted some leeway by zoning and code officials, this is not an issue that is best sprung on them by a contractor during permit applications or site mobilization by a contractor.

For these reasons, the following is recommended to civil engineers and other design professionals during the preparation of the construction documents:

STOCKPILED MATERIALS

Stockpiled materials come in a variety of types. Contractors who process material on-site for reuse as mulch, aggregate, or topsoil may need to store this material for much of the construction phase until it can be used for paving, final grading, or landscaping near the end of the project.

Recycled material stockpiles can easily erode due to wind or water action, and this erosion can contribute suspended solids, nutrients, metals, and harmful pH into stormwater systems or natural habitats. Protecting stockpiled materials from erosion is necessary, and required by ordinance in many locales. The following practices are recommended for any contractor engaged in storage or stockpiling of erodible material on a long-term or temporary basis.

1 Cover and contain the stockpiles of recycled material to prevent stormwater from washing material to surface waters or a storm drainage system.

2 Ensure that stockpiled material areas are not in a natural drainage path, or are protected by silt and erosion control barriers.

3 The covers must be in place at all times when the stockpile is not in active use.

4 Do not hose down the stockpile area to the storm drainage system.

5 Locate stockpile areas away from paved areas.

6 If the stockpiles are so large that they cannot be easily covered, use silt fences and other erosion control practices at the perimeter of the stockpile area to prevent runoff from the site or into the storm drainage system.

STORMWATER RUNOFF, EROSION, AND SEDIMENTATION CONTROL

Soil erosion is defined as the wearing away of land by the action of wind, water, gravity, or a combination of all three factors. The result of this erosion of land is that soil particles are dislodged and moved to other locations. Soil particles that have been transported to new locations and deposited are called sediment. Sedimentation can adversely affect plant and animal environments. The best way to combat sedimentation is to control the erosion of soil particles from construction sites. Construction projects often require the removal of layers of vegetative cover in order to clear area to work on the site, to complete the construction of foundations, and to accomplish the final grading, paving, and landscaping required in the construction documents. All of this work carries with it the possibility of erosion. With proper design considerations and site management during construction, soil erosion can be minimized and sediment can be controlled so it does not leave the site.

Techniques used to control soil erosion and sedimentation control include:

- Covering stockpiled materials
- Hay bale dikes
- Silt fences
- Temporary grading (slope diversion or diversion dike)
- Sediment traps and basins
- Stabilized construction entrances and drives
- Inlet and outlet riprap
- Water bars (a ridge or channel constructed diagonally across a road or ditch)

Nationwide, soil erosion districts and municipalities require separate document filings and permits before a construction project can begin. The plans are typically prepared by the civil engineer hired by the owner or architect, and are sometimes filed months before the contractor is even selected. Contractors implementing a recycling plan often find they must amend the permit, or work with the district inspector, to manage the stockpiling of building materials on-site. Soil erosion and sedimentation control plans also do not always reflect the real size of recycling zones on the project, or account for access to this zone from the road or points on a large site.

11

MARKETING CONSTRUCTION SITE
RECYCLING

Recycling Marketing and Public Relations

A contractor's marketing of his jobsite recycling program has two goals. The first is to help locate markets for the recycled waste from the project and the second is to gain publicity for the firm by touting their "greenness." Marketing to potential recyclers and public relations would normally be considered separate endeavors. In sophisticated marketing operations, each would be on a separate track and involve individual strategies, goals, and budgets. Most contractors are not interested in, or capable of, mounting separate marketing and public relations campaigns for an individual project, no matter how large. For this reason, we suggest that contractors interested in promoting their recycling efforts on a specific project use a combined effort that uses simple local outlets and tools to accomplish both goals. Both marketing and public relations are worthwhile objectives, but marketing is a dicey business, and publicity can be a two-edged sword. Here are some of the pros and cons of marketing recycling:

Pros

- It may put the contractor in contact with lower cost markets and haulers.
- It may garner good public relations and all that comes with it, including additional business and an enhanced reputation.
- It may enable the contractor to recycle odd products that as yet have no established market.
- It may attract more architectural salvage bidders or opportunities to the project.

Cons

- It alerts illegal dumpers to the availability of containers and their location.
- It alerts metal thieves to the possibility of profit and its location.
- It may generate contacts with some individuals and firms who have no ability to perform the work, and divert valuable staff time in responding to these "dead-end" inquiries.

Whether the pros of advertising outweigh the cons depends on the particular contractor and project. When a contractor faces an ambitious recycling rate requirement on a project, however, he likely needs to seek every advantage he can find and scour the area for potential markets for his waste. Marketing can help in that regard, but it will never beat the old-fashioned deskwork of digging through databases and phone books and talking with locals in the industry.

The good will that accompanies recycling efforts in construction deserves special mention. Demolition contractors were once considered the cavemen of the contracting industry. They were not recognized as craftsman in the mold of other tradesmen on the construction site, and not held in high esteem by either the public or the contracting industry. Recycling of C&D waste has dramatically changed the perception of the demolition industry, and the recycling operations on many projects—often led by the demolition contractor—have burnished the reputations of the demolition and construction industries alike. The public at large, and business owners in particular, more fully support recycling programs and are willing to recognize and be associated with any contractor who is willing to promote his recycling efforts.

For example, The Cherry Companies of Houston, TX, a group of integrated demolition and waste recycling companies were recognized as the 2009 winner of the Ernst & Young Entrepreneur of the Year award in the construction and industrial services sector for the Gulf Coast area.[1] This type of recognition typifies the new respect waste recycling is garnering for the industry.

Launching a basic marketing campaign need not be a daunting task, even for the overworked back-office staff of a normal general or demolition contractor.

Marketing of a C&D recycling effort would normally occur both before and after the actual operations on the site. Before the start of construction, the contractor can market to identify potential markets, stir up some additional competition among haulers and recyclers to lower his costs, and let the surrounding communities know that the firm is an active recycler. If the recycling plan is complete, and the recycling rate is already defined, he may even wish to announce it in press releases or advertising: "Acme Construction will be recycling at least 75 percent of the construction and demolition waste from the jobsite."

After construction is complete, the contractor should use local advertising to announce the recycling achievements of the project (the actual recycling rate—assuming it exceeded the goals announced preconstruction) as well as thanks for the recyclers involved with the project. A savvy contractor will also prominently mention the owner and design professionals, since this goodwill gesture may help in generating future work.

A *typical preconstruction* marketing effort could consist of the following elements:

■ Advertisement in the local newspaper, or even an RFP announcement in the paper's legal notices section. Because of the inherent goodwill recycling programs generate, it may even be possible to garner some free publicity in the form of a news article, though this could be counterproductive to the contractor's goal of finding recyclers with experience in the industry, and not just the time to read the newspaper and figure themselves qualified.
■ Contacts and postings at the local builder's exchange, union halls, or chamber of commerce.
■ Temporary site signage or banners.
■ Press releases to local newspapers announcing the recycling program.
■ Mention of the recycling endeavor on the contractor's or project owner's Web site, local community Web site, or a dedicated Web site set up expressly for the project.

For projects located in residential areas, or with housing near the construction site, some specifically conceived neighbor relations also can be beneficial. Personal contact with individual neighbors, a public question-and-answer meeting with a neighborhood group, or targeted notification to the affected neighborhood can go a long way toward helping to dampen anger at dust, noise, and traffic inconveniences generated by construction activities. People tend to be more understanding when they are forewarned, when they understand the inconvenience is short-term, and when they are aware that the project has among its important components the good cause of recycling.

For instance, careful phrasing in a community flyer can help the contractor negotiate the shoals of public inconvenience. Some examples of flyer statements:

■ *Acme Construction's on-site recycling operations are a part of our commitment to sustainable construction. They may generate noise and dust for short periods during the course of the project. Although we will use measures to limit the inconvenience for our neighbors, we ask for your understanding during the 5 months the project will be underway.*
■ *Acme Construction is proud to announce that we will be recycling at least 75 percent of the construction and demolition waste from the Park Center Development. Although we will do our best to limit inconvenience to our neighbors, some of our recycling efforts will generate noise and dust during business hours as we work to remake debris into products that can be reused in other ways. We will work to minimize this inconvenience, and appreciate your understanding during the 3-month period when recycling operations will be occurring.*

In communicating with the public, contractors must be careful to honestly state the types of disruption that may occur and the anticipated duration. Always use the word "recycle," but do not attribute all site noise, traffic, and dust disruptions to recycling efforts when other causes are to blame. The public perception of the value of recycling will help to make neighbors more understanding of any inconvenience it causes, but using this benefit to explain away the inconveniences caused by normal construction activities should be avoided.

Various recycling consultants, who normally provide marketing assistance to contractors in identifying potential local markets, may also have some expertise in assisting a contractor with reaping some public relations benefits from his recycling effort; however, this latter goal is best achieved by a public-relations consultant. Contractors must take particular care to avoid the practice of *greenwashing*, or making environmental or recycling claims that are only partially true or cannot be supported by data. In terms of project waste management, such claims could take the form of claiming untrue recycling rates or making claims for the end-use benefits of recycled products that cannot be verified or are speculative at best. The real benefits of C&D recycling are real and substantial. Contractors should claim the public relations benefits of their recycling work, but avoid the temptation to elaborate on them.

PUBLIC RELATIONS STRATEGIES

The most experienced organizations in marketing recycling C&D efforts are local governments. They long ago recognized the value of reducing both the municipal and construction waste streams as both an economic and environmental savings for their communities. As a result, they were the leaders in creating public relations campaigns and guides for consumers to follow in reducing household waste. Many cities have also developed C&D waste management guides for contractors, to promote jobsite recycling. Although contractors experienced in recycling will find most of these guides to be simplistic, many guides do offer a list of construction waste recycling markets in the area.

Contacting the local recycling or municipal waste office is an excellent first step for a contractor seeking to market his waste to the local recycling market. Local authorities may be aware of additional markets not yet published in their directory, and can be invaluable in assisting a contractor in dealing with various types of hazardous waste (and universal wastes) he may encounter in his project.

Because they possess this recycling promotional expertise, and often have well-known Web sites and publications developed to support it, municipalities can be an ideal way for contractors to claim some public relations goodwill from their recycling program. Mecklenburg County, North Carolina's *Wipe Out Waste* program, for example, provides free business recognition in the local media to contractors with C&D waste programs underway in their area. King County (Seattle), Washington's *Green Tools* online newsletter provides free announcements of business recycling efforts, including C&D recycling. Announcements from government entities also carry the value of impartiality, a valuable commodity in public relations.

Steve Bolerjack, communications director for the Five O'Clock Club, a career counseling network, offers the following tips for small business public relations: define the audience, develop a public relations plan, create a strategy, develop tactics, and cultivate media relationships. Although Bolerjack was offering general guidance for small businesses in general, the examples below apply to demolition and general contractors seeking to promote their recycling efforts:

1 *Define the audience*: Contractors need to define whether they are seeking to publicize their recycling program with potential business owners (future clients), potential markets for recycled products, or the public at large.
2 *Develop the public relations plan*: Identify the key goals, decide how the firm wants to be perceived (positioning), and define the key facts that need to be communicated.
3 *Create a strategy*: What is the message that needs to be communicated to each audience? Is the goal, for instance, to inform potential corporate clients that the contractor is very experienced in saving owners' money through recycling C&D waste?
4 *Develop the tactics*: How can the firm best communicate these facts to the audience? If the goal is to communicate the cost savings to owners through a contractor's expertise in recycling, what specific means can best accomplish this goal? Is it through reprinting and distributing favorable local press or municipal announcements? Or is the best method the creation of a flyer or white paper with factual information about actual cost savings to owners resulting from the company's recycling proficiency?
5 *Develop a relationship with the local media*: Make frequent announcements to the local media regarding meaningful information about the business, including major personnel changes, accomplishments, awards, and completions. Try to make the job of local reporters and editors easier by preparing professionally written press releases, clear black-and-white photography, and unusual or fresh angles that would be of interest to the media's audiences.

Bolerjack concludes with this simple advice for small-company public relations: "Solid public relations include community participation, bylined articles, public speaking, media commentary, relationships with local reporters, and development of good professional citizenship."[2]

Marketing to the Public

Some construction companies have used direct public marketing in recycling materials from their sites. This would appear to be most practical on projects with these characteristics:

1 The project is located in areas with very poorly developed recycling markets.
2 The project has limited demolition, with quantities that are too small to interest local recyclers.
3 The nature of the architectural salvage is such that salvage firms are not interested in the project, but individual homeowners or businesses may be.

In such circumstances, the contractor becomes a recycling retailer, offering his limited recycling goods directly to the public. Though necessary on projects where there are no other options, directly public marketing of recycled products carries some hazards for the contractor. Even if the amounts of recycled material are limited, marketing in this fashion is time-consuming and labor intensive. For reduced liability, the contractor would be well advised to remove the materials from the facility with professional demolition forces and not allow the public into the work zone. While she can easily document the sale (or more likely, the free acquisition) of the material by individuals, it is a much more difficult task to obtain assurance that the material, or some percentage of it, does not ultimately end up in a landfill. Municipalities or owners may not accept public marketing bills of sale or transfer forms as evidence that the material in question has, in fact, been recycled. For this reason, direct public marketing should be viewed as a limited public relations adjunct to a standard recycling program rather than the heart of the program.

Given those caveats, following are some means of marketing recycled materials directly to the public:

- Post signs offering free materials to the general public.
- Advertise reusable items in the newspaper.
- List items in local materials exchange Web sites or publications.
- Arrange for materials dealers to collect materials directly from the jobsite.
- Conduct a site sale (a yard sale of sorts) at the job site to sell reusable items.
- Allow employees and subcontractors to remove materials for their own use.

PRESS RELEASE TIPS

The press release is the basic implement in the marketer's toolkit of ways to spread the news about the contractor's recycling products, progress, or success. Press releases can be used to alert local recyclers about the general facts related to the project recycling efforts (types of material, when available, whom to contact). A preconstruction press release can also be used to simply tout the contractor's intent to recycle and the goal he will achieve. After the project is complete, a follow-up press release may document the total recycling plan accomplishments, perhaps including some basic statistics about the amount of tons of material diverted from the local landfill, or the potential end uses for some of material recycled from the project.

See Fig. 11.1 for tips on preparing effective press releases.

For this good news to reach the public, the contractor or his public relations representative, must write a press release that is sufficiently interesting or compelling to persuade local newspaper or web editors to publish information derived from the release. Editors of most newspapers must sift through a regular stream of press releases of all types from a variety of organizations. A contractor's press release will be vying against those from other organizations, including nonprofits and charities, with perhaps more compelling messages and needs. To be successful in shining through this pile of other press releases, here are some guidelines

- Cover the basics: Include contact information and clear heading.
- Make it compelling: Create an interesting lead or hook.
- Make it clear: Keep the writing simple and straightforward.
- Make it interesting: Use quotes and interesting facts.
- Make it accurate: Use verifiable facts.

Figure 11.1   Press release tips.

for creating a compelling press release that avoids the discard pile or the delete button:

Cover the Basics   Develop a letterhead template that includes the words "News Release" or "Press Release" at the top. This letterhead should also include the name of the contractor's key contact person, phone number, fax number, and e-mail address.

Create a Lead or Hook   Create an opening sentence with a news angle to hook to interest the editor and keep him reading. This hook, ideally, should include the contractor's or project name and an action that has or will occur.

Keep it Simple and Straightforward   Avoid grandiose language, flowery prose, and overuse of adjectives. Use clear and direct language with action verbs and specific information about the recycling program. Vague and self-laudatory language reads false in press releases, and causes editors to wonder if the press release is seeking to create good news that has not been fully earned.

Make it Interesting   Quotes and interesting facts give life to the press release. Include plenty of quotes, because these are usually what journalists pick up from releases. A quote from the owner, public official, or person outside of the company issuing the release is particularly compelling.

Use Verifiable Information   Never use statements in press releases that cannot be documented or verified by others. Journalists may ask for additional information to support claims in a press release, especially if the release has been successful in creating enough interest that the editor would like to run a story on the project. Good media relations are based on trust and accuracy. Inaccurate (or deliberately false) information will ruin a company's reputation with the area press and make future public relations efforts much more difficult.

Marketing to Recyclers

The recycling markets in most areas, particularly urban environments, are composed of professional and savvy businesspeople. They are "markets" in only the broadest form of the word, since mostly what they do is simply accept recycled material. Only in the cases of architectural salvage, metals, and a small and select number of other recycled materials is the contractor likely to reap payment from the recycler. In all other cases, he will pay the recycler to accept his debris, or the recycler will accept it at no cost, with the contractor either paying the cost of storage or hauling or not. The contractor benefits in diverting material to recycling markets and meeting the municipal or contractual recycling rate requirements, and will also reap some financial benefits in that recycling does save money over the costs of landfilling all the construction and demolition waste generated by a project.

Given these realities of the recycling marketplace, how does a contractor go about soliciting competitive "bids" from a market where the primary objective is not to maximize income, but to minimize costs?

The ideal method is to approach these markets in the same manner a contractor would any other subcontractor or supplier: Define the work and obtain a proposal. The vagaries of construction are challenging enough without confusing matters further by approaching markets haphazardly. This is even truer in the realm of recycling, where recycler capabilities and business practices can vary widely. Consider:

1 Some large recyclers can handle a range of debris of different types, and can provide their own containers and hauling.
2 Small recyclers may pay a higher rate for high-value metals, but the contractor must pay for his own containers and haul the material to the recycling facility.
3 Some recyclers accept uncrushed concrete waste; others will only accept crushed material. As with all other forms of subcontracting, weighing proposals from recyclers requires a careful assessment of what is—and is not—included in the agreement. Experienced contractors also know to use a skeptical eye in reviewing proposals from recyclers who agree to regular hauling schedules but do not clearly have the means to meet such a schedule. Agreements can stipulate all manner of requirements, deadlines, and damages, but once the project is underway and the recycler is not performing as promised, there is little the contractor can do other than complain loudly and try to work through the situation.

Homework in investigating the recycler's capabilities, therefore, is essential on the front end before signing the contract.

Approaching recycler firms (marketing to the markets) should include the following key elements:

Define the Products   It is critical to define the debris being offered and its estimated quantities. The amount of material may influence the value to a particular recycler, and determine whether he is willing to make containers and hauling available to a contractor and at what cost. A contractor approaching recycling markets for the first time, and inexperienced in what types of waste they accept, may need to hire a consultant to assist him. It is appropriate to list all the recycled waste on a single form for use by all recyclers. The benefit in this method is that it allows large recyclers who accept a range of different waste products to combine their offer to the contractor into a bulk rate. As noted previously, it is much to the contractor's advantage to deal with as few recyclers as possible in terms of minimizing his management and administrative time. Similarly, recycling is a bulk business. A recycler would prefer to commit his storage containers and hauling capacity to one job if it will yield enough waste material to warrant the expense. In that sense, it is mutually helpful for large recyclers and large contractors to seek each other out, and the contractor should provide such a recycler with a proposal form that makes it easy for him to offer the contractor the benefits of his size and capabilities.

The only situation in which a contractor would not want to carefully define the scope of products is in the realm of architectural salvage. Salvagers should be asked to survey the building and specifically list all items they will salvage, and the amount they will pay for the specified products. If they are not offering to pay for their salvage, or are intending to charge the contractor, he must carefully consider whether it is better to include these items in the general demolition contract as recycled waste.

Provide a Standard Form   The contractor should ask each recycler to complete a standard proposal form (see this book's Online Resources for a sample form). This form should list recycled waste available, the estimated quantities, and include any stipulations or contractual details the contractor feels are critical to his needs. In general, such stipulations should be limited since, with a few exceptions, recycling is largely a buyer's market. Recyclers reading a long list of conditions attached to a proposal form may well decide dealing with a particular contractor will be more trouble than obtaining those products would be worth. In the high-volume, low-margin world of recycling, a contractor with such a stance may find himself with an insufficient number of markets for his C&D waste.

The form should also include a few other basic questions:

- *Hauling*: Will the contractor provide hauling from the jobsite to his facility? If so, how often?
- *Containers*: Will the contractor provide containers? If so, what size and how many?
- *References*: Ask for three contractor references.
- *Proposal price*: What will the contractor pay or charge for each unit of each product?
- *Documentation*: What documentation form will the contractor provide for each load? Is end-user documentation available?

Assess Time and Money in Weighing Proposals   Management time is money, and a recycling proposal that includes containers and hauling reduces the management burden on the field superintendent and project manager. That fact should be worth money to the contractor, although its value is difficult to quantify. Similarly, dealing with a single recycler for a number of waste products should make the management of the recycling plan easier, and allow more time for the field superintendent to be concentrating on the area where a contractor truly makes his money—the construction of the project. All of this argues for looking beyond the simple dollar costs of recycling proposals to the larger goal of meeting the recycling rates with as little difficulty as possible.

Include the Recycler in Planning the Work   The recycler, like any other sub-contractor, is an expert in his area. The contractor should include him in the detailed planning of laying out the recycling zone on the jobsite, and determining the shifts of container sizes and types as the demolition and construction moves forward.

Recycling of Unusual Materials

Contractors often inherit demolition sites containing a wide variety of debris. Vacant industrial and commercial sites sometimes become the community's dumping ground, containing everything from abandoned vehicles and trashed furniture to drums of unknown chemicals. Where hazardous waste is suspected, refer to the Hazardous Materials section in Chap. 2 for tips on how a contractor can protect himself from accepting unlimited liability for hazards found on the jobsite. The unusual variety of waste found on such a site presents equal challenges for the recycling contractor. Because some of these products are not appropriate for mixed waste, the contractor must decide between investing the time, which may be wasted, in finding a market for his unusual waste, or simply disposing of it. There may be a market for unique jobsite waste, but finding and managing the market may simply not be worth the contractor's time and resources. This is a judgment call based on the individual situation and the amount of waste available. Finding markets for unusual products, however, is an endeavor that requires some staff time in scouring local yellow pages, online contractor resources, and conversations with area demolition and recycling contractors. Products that are not recyclable locally may be accepted by national or regional companies, though the cost and trouble of transporting the products to such companies may be not be worth the landfill savings. Still, on those projects where a contractor is obligated to meet an exceptionally high recycling rate, effectively recycling deposited waste on a jobsite could make the difference between in meeting the rate goal or not.

Following is some information relating to types of unusual waste and some common national markets available for them:

Aluminum Foils and Laminates

■ Connecticut Metal Industries: www.foilfoil.com

Antifreeze (Ethylene Glycol, Propylene Glycol)   Do not mix antifreeze with other materials such as oil, gasoline, water, or hazardous waste. Recycling methods include reclamation or reuse, but not burning for energy recovery.

■ Republic Services offers antifreeze recycling in many areas they serve: www.republicservices.com

Auto/Truck Plastic Body Parts

■ Green Earth Plastix: Gainesville, GA. 770.540.9358, www.greenearthplastix.com

Batteries (Consumer Electronics)

■ Rechargeable batteries of all kinds: http://www.rbrc.org
■ Some Best Buy stores accept rechargeable batteries
■ Some Staples office supply stores accept rechargeable batteries

Bed Liners (from Pickup Trucks)   United Plastic Recycling, Atlanta, GA. Contact: Parker Pruett, 404.351.5033

Books (Hardbacks, Softbacks, and Manuals)

■ Recyclable as mixed paper waste, but if a large quantity exists on a site it may be desirable to contact a specialty recycler
■ Caraustar Recovered Fibers: 770.451.3271, www.book-destruction.com/

Boxes (Corrugated Reusable)   As discussed in Chap. 3, corrugated cardboard (OCC) is readily recyclable as mixed paper waste. This section refers to specialty boxes, or corrugated containers that may have reuse value. Note that paper and waxed corrugated boxes cannot be recycled with mixed paper waste. They are a specialty item, and require identifying a separate recycling market.

■ Rebox Corporation's Web site provides a market to sell used corrugated boxes: www.reboxcorp.com
■ U-Haul Box Exchange is a message board for trading, selling, or buying reusable boxes and moving supplies. www.uhaul.com/boards/

Building Materials (Surplus, Salvaged)   Wood and interior finish products that lack sufficient quality to be of interest to architectural salvagers may be of value to nonprofit affordable housing organizations in the area. The largest national organization is Habitat for Humanity.

■ Look for Habitat ReStores, located throughout the country.

Bumpers (Plastic, From Cars/Light Trucks)

■ Keystone Automotive Industries: 404.691.6930. Will buy only repairable units from model year 2000 or newer, and will pick up no charge (dependent on area)
■ BumperPlus will pay modest sums for repairable plastic bumper covers. 770.422.7587, www.bumperplus.com

Cable Reels and Spools

- Baker Reels: Hartselle, AL 35640. 800.633.3962
- Carris Reels: www.carris.com
- Sunoco: www.sonoco.com/sonoco/Services/Recycling/rls_reel_recyclng.htm

Cds, Dvds, Audio/Video Magnetic Recording Tapes, and Data Cartridges

- http://www.grahammagnetics.com/surplus_form.html
- http://www.semshred.com/
- http://www.lacerta.com/
- http://cdrecyclingcenter.com/

Clothing and Textiles

- http://www.leighfibers.com/
- http://www.dumontexport.com/index.html

Coin Recycling (Bent, Damaged, Discolored)

- Superintendent, U.S. Mint, P.O. Box 400, Philadelphia PA 19105. 215.408.0205

Cooking/Frying Oils (Used)

- Location dependent. Investigate local markets.

Drums (Plastic, Steel, Fiber)

- **Container Depot, Atlanta, GA. 404.863.1257. Charity America** connects donors with charities across the nation.

Electronic Waste

- *Computers for Schools*: An association that reconditions donated computer equipment for educational use.
- *Goodwill Industries*: Many Goodwill Industries locations accept donations of newer televisions and computers.
- *National Cristina Foundation*: Accepts donated computers to give to people with disabilities and others at risk.
- *Apple*: Mail-in service for cell phones and iPods.
- *AT&T Wireless*: Some locations accept cell phones and PDAs, plus accessories and batteries for donated devices. In-store drop-off (AT&T-operated and participating authorized dealer stores only).

- *Best Buy*: All stores accept desktop and notebook computers and peripherals, DVD and VCR players, small electronics, telephones, and televisions and monitors up to 32 in. A $10 fee per unit recycling fee is charged for items with screens, such as televisions, laptop computers and monitors, but the consumer instantly receives a $10 Best Buy gift card in exchange for the recycling fee. Company also offers recycling grants to municipalities and nonprofit organizations.
- *Costco*: Computers, CRT and LCD monitors, camcorders, digital cameras, fax machines, game systems, MP3 players, PDAs, personal printers, and smart phones. Free mail-in recycling. Trade-in value paid in Costco gift cards for some items. Costco web site
- *Dell*: Computers and peripherals. Free recycling of Dell-branded products at any time and of any brand when a consumer purchases new Dell equipment.
- *Hewlett Packard (HP)*: Cell phones, computer hardware, inkjet and laser printer cartridges, rechargeable batteries, and user-replaceable mercury-added lamp assemblies. Drop-off and mail-in.
- *Lexmark: Printers, inkjet cartridges, and toner cartridges. Mostly mail-in.*
- *LG Electronics (includes LG, GoldStar, & Zenith)*: Most LG, GoldStar, and Zenith products, except cell phones. Consumers can recycle up to five units per day, free of charge, by dropping them off at designated Waste Management Inc. (WM) eCycling locations. Cell phones are covered under a separate program.
- *Office Depot*: Cell phones, computers and peripherals, cords and cables, CRT and LCD monitors, digital and video cameras, DVD and MP3 players, fax machines, inkjet and toner cartridges, pagers, PDAs, printers, rechargeable batteries, small TVs, telephones, and VCRs. In-store drop-off.
- *Radio Shack*: Many but not all brands accepted. Cameras, camcorders, cell phones, GPS systems, laptop computers, MP3 players, and video game consoles. Mail-in item(s) and receive a Radio Shack gift card based on condition and value of trade-in(s).
- *Samsung*: Samsung accepts donations of all used Samsung products. Free recycling at drop-off locations and events sponsored by Samsung, retailers, and recyclers. Additional information is available on their corporate Web site.
- *Sony*: All Sony products. Consumers can recycle up to 5 units per day, free of charge, by dropping them off at designated Waste Management Inc. (WM) eCycling locations.
- *Staples*: Cell phones, inkjet and toner cartridges, pagers, PDAs, and rechargeable batteries. In-store drop-off.
- *Sprint*: Cell phones, accessories, batteries, and connection cards that are no longer being used. In-store drop-off and mail-in.
- *T-Mobile*: Accept cell phones, accessories, batteries, and PDAs. Allows in-store drop-off and mail-in.

Expanded Polystyrene (EPS, Styrofoam in Compacted Truck Loads)

- FP International: 732.736.0337.
- Recycle-Tech. Corp. www.recycletechno.com, 201.294.8942.

Fiberglass (FRP = Fiberglass Reinforced Plastic)

- http://compositeworld.com/zk/recycle.shtml

Flexographic Printing Solvents

- National Solvent Exchange: 404.605.0085

Fluorescent Lamps (HID Lamps, Thermostats, Thermometers, and Any Devices Containing Mercury)

- www.aercrecycling.com/
- www.recyclebulbs.com/

Fog (Fats, Oils, Grease)

- Atlanta Community Food Bank, Atlanta, GA. www.acfb.org, 404.892.3333

Freon™ and Other Refrigerants

- CFC Refimax, 770.984.2292.
- Clean Air Refrigerant Recovery: 800.561.2915.

Furniture and Home Furnishings (Household Only)

- Furniture Bank of Metro Atlanta. 404.355.8530, http://www.furniturebankatlanta.org/, info@furniturebankatlanta.org

Glass (Bottles, Plate, Window, Auto, Borosilicate)

- http://strategicmaterials.com/index.html, College Park, MD. Contact: Hazel Mobley, 404.349.6952.

Halon (Was Used in Fire Extinguishers Until EPA Banned it in 1996)

- www.halonbankingsystems.com/services.html, 800.840.7698
- www.ushalonbank.com, 732.381.0600

Latex Paint

- SYNTA, Inc., Decatur, GA 30030. 404.373.7284

Lead Aprons (for Radiation Protection)

- www.shieldingintl.com

Mercury-Containing Devices (Thermostats, Thermometers, Light Bulbs, Switches)

- www.lightbulbrecycling.com
- *Bethlehem Apparatus, Inc.*: www.bethlehemapparatus.com/mercury-spill-kits.html

Microfiche Recycling or Destruction

- www.prismintl.org

Old Diesel Fuel, Stabilizers, and Restorers

- www.batterystuff.com/tutorial_fuel_storage.html
- www.solareagle.com/pri.html

Pallets and Wooden Shipping Containers

- www.palletconsultants.com/PurchaseAndRemoval.php

Plastic Auto/Truck Parts

- Green Earth Plastix, Gainesville GA. 770.540.9358, info@greenearthplastix.com

Plastics   Specialized buyers of specific resins:

- #1 PET: Buyer—Mohawk Industries: 706.856.6481
- #2 HDPE: Buyer—KW Recycling: 800.633.8744
- #3 PVC: Buyer—Somerset Recycling: 606.274.4170
- #4 Vinyl siding: Buyer—Reilly Recovery Systems: 919.933.3611
- #5 LDPE (stretch wrap, bags, films, etc.): Buyer—Conex Plastic Industries: 706.453.2303.
- #6 PP: Buyer—Waste Stream Solutions: 540.344.3600, www.fibcrecycling.com/
- #7 PS (Expanded Polystyrene, "Styrofoam"): See above, EPS

Polyurethane (Rigid, Foam, Flexible)

- www.polyurethane.org/

Porcelain Products (Bathroom and Kitchen Fixtures, Crushing for Recycling)

- Check with local concrete recyclers. Porcelain can be recycled into road aggregate or aggregate for cement mixes.

Powder Coat Paints

■ Surplus Coatings, Inc., Grandville MI. 800.804.8003, www.surpluscoatings.com

Prescription Drugs (EPA Approved Disposal Methods)

■ http://whitehousedrugpolicy.gov/publications/pdf/prescrip_disposal.pdf

Printer cartridges (Laser Jet, Ink Jet)

■ Many Staples and Best Buy stores accept cartridges for recycling.
■ Ricoh products only: http://www.ricoh-usa.com/about/environment/manage_recycle_toner.asp.

Propane Tanks (Recycle 20-lb Size and/or Buy a New Full Tank)

■ http://www.bluerhino.com/BRWEB/Tank-Exchange.aspx

Scrap Tires (Auto, Light Truck, Semitrailers)

■ Contact area Portland cement manufacturers. Scrap tires can be used to fire cement kilns.
■ Rubber Manufacturers Association: www.rma.org
■ U.S. Environmental Protection Agency: www.epa.gov/epawaste/conserve/materials/tires/tires.pdf.

Shipping Containers (Barrels, Drums, Bins, Pails, Totes, etc.)

■ www.reusablepackaging.org

Solvents

■ http://solvdb.ncms.org/savingsdata.htm

Super Sacks/Bulk Bags

■ http://www.fibcrecycling.com/

Teflon™

■ www.reprolontexas.com/products.htm, Burnet TX. 800.846.6201
■ www.csiplastics.com, Holden MA. 508.829.7727

Textiles, Fibers, Cloth

- www.textilefiberspace.com/a/tx1400.html
- www.leighfibers.com/

Thermostats, Old, Containing Mercury, Intact

- Lennox Commercial, 1650 Indian Brook Way, Suite100, Norcross, GA. 770.925.8060

Transparencies (for Overhead Projectors)

- http://solutions.3m.com/en_US/

Tyvek™ (Envelopes, Clothing)

- http://www2.dupont.com/Tyvek_Envelopes/en_US/tech_info/tech_environ.html
- www.garmentrecoverysystems.com/

Waxed Corrugated Boxes

- Enviro-Log, Inc., www.enviro-log.net. Contact: Ross McRoy, 866.343.6846

X-Ray Films (Exposed, Medical Or Industrial) for Silver Recovery

- United DMS of Tennessee: www.uniteddms.com
- www.promedrecycling.com

References

1. Taylor, Brian. "Built on Recycling." *Construction & Demolition Reycling magazine* September/October 2009: 20-22.
2. Bolerjack, Steve. "Public Relations Tips for your Small Business." The Five O'Clock Club. November 27, 2009 <http://www.fiveoclockclub.com/articles/1997/10-97-BolerjackPR.html>.

12

RESOURCES

See Fig. 12.1 for a list of additional online resources on the McGraw-Hill Web site.

Contaminants in C&D Waste

- Acetone
- Acetylene gas
- Adhesives
- Ammonia
- Antifreeze
- Asphalt
- Asbestos (friable and nonfriable)
- Benzene
- Bleaching agents
- Carbon black
- Carbon dioxide (in cylinders)
- Caulking, sealant agents
- Caustic soda (sodium hydroxide)
- Chromate salts
- Chromium
- Cleaning agents
- Coal tar pitch
- Coatings
- Cobalt
- Concrete curing compounds
- Creosote
- Cutting oil
- De-emulsifier for oil
- Diesel fuel
- Diesel lube oil
- Etching agents

Tip Box

McGraw-Hill Construction's Web site contains a number of resources developed in conjunction with this publication. They can be accessed at: www.mhprofessional.com/rcadw. Online resources include:

■ Sample Long- and Short-Form Waste Management Specifications

■ Sample Long- and Short-Form C&D Waste Management Recycling Plans

■ Recycling Plan Calculators

■ Recycling Zone Signage Templates

■ Miscellaneous Recycling Document Templates

Figure 12.1   McGraw-Hill online resources.

■ Ethyl alcohol
■ Fiberglass, mineral wool
■ Foam insulation
■ Freon
■ Gasoline
■ Glues
■ Greases
■ Helium (in cylinders)
■ Hydraulic brake fluid
■ Hydrochloric acid
■ Insulation
■ Kerosene
■ Lime
■ Lubricating oils
■ Lye
■ Methyl ethyl ketone
■ Motor oil additives
■ Paint/lacquers
■ Paint remover
■ Paint stripper
■ Particle board
■ Pentachlorophenol
■ Polishes for metal floors
■ Proportion of concrete in various flooring systems
■ Putty

- Resins, epoxies
- Sealers
- Shellac
- Solder flux
- Solder, lead
- Solder, other
- Solvents
- Sulfuric acid
- Transite pipe
- Varnishes
- Waterproofing agents
- Wood preservatives

Equipment

A wide variety of equipment is available to contractors, often through rental, to assist them in more efficiently recycling products on the jobsite. Contractors need to make the judgment early in the project whether the equipment cost and labor will pay off in reduced recycling or transportation costs. Heavy, bulk items such as concrete, masonry, asphalt, and wood and metal scraps are the best candidates for on-site processing, though the economics are not consistent from project to project. Where a concrete recycler is located relatively close to a project site, and has his own fleet of trucks, it will likely work out much better for the contractor to negotiate with the recycler to pick up the bulk materials on the site rather than attempt to process them to save on transportation costs. In the case of a bulk aggregate buyer who is located some distance from the site and has no capability to mill masonry or concrete for his mix, the contractor may find it more advantageous to grind the material on-site to be able to recycle to a particular company and minimize hauling costs.

BULK REDUCTION

Various mechanical methods are available to contractors and/or recyclers seeking to reduce the bulk of recyclable waste. The most commonly used machinery falls into the following categories:

Mobile Bulk Reduction Attachments Rolling stock equipment can be fitted with various attachments for reducing the bulk of large wood waste materials. The types of attachments available for front-end loaders and excavators include shears, pulverizers, densifier grapples, and tree stump splitters. These attachments can be customized to reduce raw materials ranging from pallets and structural framing to tree stumps.

Rolling Stock Compaction This method of bulk reduction is accomplished by driving the rolling stock over large wood waste before loading it into the processing system. This type of compaction may create too many small particles and can introduce contaminants. Generally, wheeled rolling stock compacts better without creating as much fines as tracked equipment.

Hydraulic Compaction   This compaction is accomplished using equipment consisting of a crushing surface driven by hydraulic cylinders. The compactor reduces bulk by compressing large wood waste against a reinforced conveyor bottom. Hydraulic compaction works well, but may require a lot of maintenance. The compactor can be a stationary hopper-fed unit that is loaded by rolling stock, or an in-line conveyor compaction unit that is operated by personnel on a picking line.

Mechanical Shearing   This equipment consists of a hydraulically driven shear that periodically slices through the wood waste traveling on a conveyor. There is a break in the conveyor bottom where the shear passes through the wood waste. Mechanical shearing can be expensive and the equipment is potentially dangerous if it is not properly designed.

Precrusher   This equipment features a rotating crushing device and a feeding mechanism to provide coarse size reduction of material prior to further screening and size reduction.

GRINDING (OR SHREDDING) EQUIPMENT

Grinders and shredders are made up of a series of blunt blades or knives fixed in rows to a rotor that passes between slots in a fixed anvil, or between fixed blades on an opposing rotor. Waste (typically wood waste) is forced through the slots by the blades and is then torn or shredded into smaller pieces. The size of these pieces can be regulated by adjusting the gap between blades. Because shredders do not require cutting or pounding, less maintenance is needed.

Shredding equipment typically operates at lower speeds (less than 20 RPM) than conventional hogging equipment, relying on high torque ratios to provide reduction forces. Hydraulic systems can apply variable force with tremendous pull-through strength, making them appropriate for mixed feed materials. High-speed shredders function more like hammer mills used in secondary size reduction. The capacity of a shredder is primarily limited only by the low speed, physical size of the shredder, and the type of wood waste being processed. Most shredders are rated for capacities by their manufacturer and may be sized to meet any volume requirement.

Shredders are capable of handling almost any type of wood waste materials, including stumps and wood resulting from land clearing, prunings, pallets and crates, construction lumber trim, panel boards, demolition waste, and secondary wood waste. Shredders are also relatively unaffected by contaminants. Small rock and metal contaminants, plastic, glass, and other foreign materials will pass through the machine without causing major maintenance problems. Operators should note that shredding high-quality, clean wood waste will not maximize the market value of the recovered material because of the shape of the shredded, finished product.

CHIPPERS

The following are the various varieties of chippers.

Disc Chipper   A disc chipper consists of a series of embedded knives arranged around a large steel disc. As the disc rotates, the knives pass a fixed anvil directly at the chipper's infeed. Disc chippers can have either vertical or horizontal feeds. The number, position, and bevel of the knives are critical to the size and quality of the chips produced, as well as to the chipping speed.

Drum Chipper   A drum chipper consists of a series of knives evenly spaced around a large rotor. The knives chip the wood waste as they pass over a steel anvil at the chipper's infeed. Drum chippers allow greater control over the sizing of the finished product. The amount of oversized chips can be controlled by placing a basket screen on the bottom of the drum. This eliminates the need to screen chips before shipping to the end user. Drum chippers produce a more consistent chip. For best results, samples of the raw material should be tested with different chippers to determine which equipment produces the most desirable results.

Types of Chipper Raw Material   Chippers are capable of processing any type of wood waste that is free of hard contaminants such as rock and metal. Since hard contaminants are difficult to restrict from many wood waste supplies, only the cleanest or most heavily sorted supplies are appropriate for chipping.

Chipper End Products   Chippers produce very high quality chips suitable for pulp and paper, and for the production of panel board.

SPECIFIC WASTE MANAGEMENT EQUIPMENT

Backhoe Loaders   A backhoe load is a box attachment for a standard backhoe that can transport bulk recycled materials across a construction site. These attachments are particularly useful to collect demolition material immediately adjacent to the work zone and transport it to the container in the recycling zone.

Balers   A specialized piece of equipment designed to convert loose waste into baled stacks of material. The benefits of baling recycled material include: occupying less space in storage and hauling and creating more profit for the recycler. Types of balers include: vertical, horizontal, textile, and foam.

Vertical Balers   Vertical balers are used to bale shredded office paper cardboard, plastic containers, steel cans, aluminum cans, vinyl siding, sheet metal, and other materials that have sufficient elasticity when stacked to allow them to be strapped with baling wire. Cardboard bales typically measure: 60 in (152 cm) long, 48 in (122 cm) wide, and as much as 36 in (91 cm) deep. They weigh approximately about 800 to

1000 lb (363 to 454 kg) each. Vertical balers are not ideal for use with materials with less compressibility such as: newsprint, office paper, and mixed paper. Many plastic recyclers and paper mills will not accept baled materials with metal strapping.

Horizontal Balers Horizontal balers are considered the most efficient type of baler. They typically produce bales that are 6 ft long. Finished bales of cardboard will normally weigh approximately 1200 lb (544 kg). These balers eject out the side or end, depending on the type of baler and space available. Horizontally baled material is usually preferred by recyclers or end users because it is consistent in density and weight, easily stacked, and takes up less warehouse area.

Breakers Breaker equipment, as the name implies, is intended to break bulk material such as concrete slabs into manageable chunks for transport or crushing, or to break recycled material such as glass into cullet sizes suitable for end-user markets. On-site breaker equipment usually consists of backhoe attachments that allow an operator to use a hydraulic, pointed ram to break up solid objects. Specialized breaker equipment for producing recycled goods suitable for markets is found at recycling centers or manufacturers who customize the equipment for their process needs.

Chippers Machines are available for use on-site that shred wood or vegetation scraps into various sizes for use as mulch or in reconstituted wood products. (See specific descriptions of chippers above.)

Container Dumpster Container dumpsters are smaller boxes that can be lifted by a hauler and emptied into a trash truck, leaving behind the dumpster for refilling. Available in a range of sizes; can be quickly emptied on-site with less area than that required for a roll-off container.

Roll-Off Containers Roll-off containers are large recycling containers transported to the site on truck trailer (Fig. 12.2). They are wined off the trailer onto a prepared pad area. They are available in a range of sizes. Though they handle more waste, they

Figure 12.2   Roll-off container. © 2010, Olivier Queinec, BigStockPhoto.com

require much greater site area for unloading, a longer transfer time, and must be replaced with an empty unit when carried off. Roll-offs are much less expensive by volume than container dumpsters.

Conveyors   Conveyor systems include a wide variety of equipment designed for both construction sites and recycling centers. Conveyors move recycled goods from a demolition or processing area on-site to a container, truck, or another piece of equipment for additional processing. Available in widths up to 6 ft, conveyors are an efficient means of raising material for dumping into a container or hopper.

Cranes   Cranes are comprised of a diverse group of machines that lift and move heavy objects. Cranes differ from hoists, elevators, and other devices intended for vertical lifting, and from conveyors, which continuously lift or carry bulk materials, in that they lift single, heavy objects one at a time. For this reason, they are less useful in recycling centers than on construction sites, where rooftop equipment and large demolished objects may need to be lifted vertically out of a structure undergoing deconstruction. Cranes found on construction sites include: fixed tower cranes, mobile cranes, truck cranes, and trailer-mounted loader cranes.

Crushing Equipment   Crushing equipment encompasses a wide range of devices intended to render bulk materials (principally concrete, stone, or masonry) into smaller goods that are easier to store and transport, or that can be reprocessed into aggregate material suitable for an end market. The most common crushers are:

Jaw Crushers   A compression type crusher used for initial crushing of larger material.

Cone Crushers   A compression type crusher used for secondary crushing of material.

Impactors   Horizontal impactors are typically a primary crusher, while vertical shaft impactors are used for making sand and other finer materials. Impactors use impact and shear to reduce material.

Roll Crushers   Produces a more consistent product with fewer fines, and works well with nonabrasive materials.

Rod and Band Balls   Use impact, attrition, and compression to reduce rock into fine particles.

Construction site crushers mostly consist of track-mounted devices with attached conveyors that are hauled to the site and used to reprocess large amounts of bulk goods into manageable sizes. In the ideal case, the crusher conveyor deposits the processed material directly into a truck for hauling to the recycling center or end user.

Dust Control   Construction dust control is desirable for a number of reasons, including: reducing wind erosion and dust, reducing respiratory problems, minimizing low visibility conditions caused by airborne dust, and reducing complaints from neighboring landowners. Dust control is mostly handled through construction jobsite practices, including an organized program of control during dust-producing activities. Portable dust and erosion control products, however, can be used by the contractor to keep airborne dust to a minimum, helping to reduce the potential for worker compensation claims and neighbor complaints. These products include soil-binding additives, erosion control mats, and other soil stabilization products.

Excavators   Excavators are any of range of hydraulically-operated heavy equipment used to dig, trench, or otherwise remove soil and stone. The two most common jobsite excavators are the backhoe and compact excavator. A backhoe consists of a digging bucket mounted in the form of a two-part articulated arm mounted on the back of a tractor or front-end loader with an open cab. A compact excavator is a track-mounted device with a two-part arm operated from an enclosed cab. The compact excavator possesses greater capacity than the backhoe, though is not as mobile. Both types of equipment can be fitted with various hydraulic attachments for different operations, such as: augers, grapples, and breakers.

Forklifts   Forklifts are propane or gas-powered devices that can be used within a facility undergoing deconstruction or at a recycling center to easily transport pallets. Forklift capacity ranges between 2500 to 5000 lb (1134 to 2268 kg), with the forklift usually weighing twice the lifting capacity. When outfitted with pneumatic tires, the lift can be used over both slabs and well-graded stone bases. When the tires have foam or rubber cores, they suffer fewer flat tires. Forklifts can be fitted with various attachments, including boxes for transporting demolition material from the work area to the site recycling zone.

Gaylord Boxes   Gaylord boxes are the workhorse of recycling, used to store and ship a wide range of recycled materials efficiently and economically. Gaylords are triple-wall corrugated containers, typically 42 in (107 cm) in all directions. They fit on standard 48 in (122 cm) by 40 in (102 cm) wood pallets, and can hold up to 3000 lb (1361 kg). They are usually shipped flat and stacked in bundles of approximately 25 pieces. They can be purchased from used box dealers and commercial recyclers.

Grinders   Grinders consist of a wide variety of specialized devices used to reduce recycled products to a smaller size. Types of products handled by grinders include wood, concrete, plastic, glass, and metal. Grinders and shredders are closely related (and considered identical to some recyclers). Grinders originated in the wood-processing industry as high-torque, low-speed processors of wood waste. Shredders are similar mechanically, but process plastic, rubber, metal, and other waste. Hammermill grinders are the traditional grinding devices often used in landscaping operations. Hammermills consisted of horizontal feed grinders and tub-type grinders, and are generally larger, noisier, and operate at a higher speed than newer single-shaft style grinders and shredders.

Figure 12.3   **Wheeled loader.** © 2010, Hamiza Bakirci, BigStockPhoto.com

Hoppers   Hoppers are rolling open-top boxes for collecting recyclables immediately adjacent to the work area and transporting them to the site's recycling zone.

Loaders   A loader is a type of heavy equipment machine that is primarily used to lift recycled material (such asphalt, demolition debris, wood) into or onto another type of machinery, which can include a feed hopper, dump truck, container, or conveyor. Front-end loaders and track loaders are the most common types of loaders found on construction sites. Front-end loaders (or bucket loaders) are wheeled loaders that are very mobile and versatile (Fig. 12.3). Track loaders are most useful in difficult terrain where sharp objects or excessive slope would make wheeled loaders difficult to manage.

Graders (Road or Motor Grader)   A grader is a construction device with a long blade used to create a flat surface. Graders are used on construction sites to create temporary or final parking areas and roads, where a relatively flat area is desired. In jobsite recycling operations, graders are not common pieces of equipment, but they are used in those situations where onsite processing results in materials suitable for use as an aggregate base. The contractor may use a grade to install the base material to get it out of the way and establish a staging area for his construction operations.

Pallet Jacks (or Pallet Trucks)   Pallet jacks are manual or electrically operated devices that allow one individual to easily move pallets of material around a plant or site. They are not typically used on construction sites since they are not suitable for any surface besides a concrete slab. Manually operated jacks are adequate. Manual jacks have capacities of 4000 to 5500 lb (1814 to 2495 kg), with 48 in (122 cm) forks that are low enough to get under standard-sized pallets.

Safety Equipment   Standard worksite safety equipment includes eye, hearing, and hand protection. Barrier devices should be employed where recycling equipment or operations pose a hazard, such as in shredding, crushing, or chipping work.

Screening Equipment   On-site screening equipment can sort aggregate sizes of crushed concrete, masonry, or stone for greater marketability and lower transportation costs.

A *flat shaker*, or vibratory, screen is used in conjunction with a stump shear to remove dirt and stones before sections of wood are processed through a hammermill grinder.

A *disc or star screen* consists of a long box containing screening discs (or stars) mounted on rolling shafts that are installed very close together across the width of the box. When used after the Hammermill grinder, prechipped material is put into one end of the box. Material of an acceptably small size falls down between the rolling shafts and onto a conveyor belt that separates it into containers or piles.

A *trommel screen* is a rotating drum screen process, in which material is put in at one end of the drum, with the acceptable sized material passing through the wire mesh on the rolling drum and falling onto a conveyor belt. Oversized material exits out the end of the drum.

Shears   Attachments for front-end loaders are available which shear metal into strips to make the material more compact and more marketable. Shears are useful in reducing the length and volume of large sheets of metal roofing, siding, gutters, flashing, and copings.

Shredders   Wood shredders are used for reducing the size of large diameter wood and stumps to a size that can be more easily handled by a secondary grinder. Shredders also loosen any dirt or rock that is attached to stumps. (See also grinders).

Skid-Steer Loaders   Skid-steer loaders of the type (commonly called *Bobcats*®) are a common feature in both construction site and recycling center operations. This device can lift approximately 1200 lb (544 kg), and is usually diesel-powered with a standard front bucket attachment. Other attachments that are useful in recycling operations include:

Grapple Bucket   This attachment allows the loader to scoop and clamp down on loose material such as paper, plastics, light metal and drywall, and cardboard.

Detachable Forks   This attachment allows the loader to double as a forklift.

Trailers   Trailers used in recycling operations range from simple open top trailers to roofed trailers with side access and interior compartments designed specifically to accommodate recycled materials. Trailer sizes vary between 12 yd^3 to 20 yd^3 (9.2 to 25.3 m^3).

Weighing Systems Weighing systems consist of on-site rented equipment that can weigh hauler vehicles before and after pickup, thereby establishing the weight of recycled material on board. Though not as preferable for documentation purposes as independent certified scales, on-site weighing systems can help the contractor verify recycling market weight statements, and allow the contractor to assess his recycling plan progress without awaiting weight statements from other sources.

Basic scale systems consist of a steel platform about 4 ft^2, connected to a digital display and a printer that can produce dated weight tickets. The weight capacity of jobsite (and recycling center) scale systems is typically between 3000 and 5000 lb (1361 and 2268 kg) with 2 lb (0.9 kg) increments. Regular calibration of the scale should be performed to ensure accuracy.

Haulers Haulers are the key to the recycling system. Without a sufficient number of hired haulers, the contractor must enter into the trucking business to carry out his recycling plan—a very disagreeable prospect. Fortunately, haulers are available in most areas, and they have an increasing familiarity with the nature of recycling hauling, including the types of containers and loads required to service the industry. Because construction hauling is a specialty, the dilemma for the contractor is to find affordable haulers with a sufficient fleet and capabilities to meet his needs. Project managers tend to find a few haulers who can meet their needs and negotiate among them for the best combination of rate and service, because the managers do not want their field superintendents otherwise needing to manage several different haulers.

The service part is important, because a lower-cost hauler who cannot pick up loads when required will cost the contractor time and money on the jobsite. A quick reference check of area haulers with other contractors should quickly reveal whether they are reliable and responsive, two keys to servicing the recycling needs on a project.

Signage Large and clear recycling zone signage is the contractor's best friend, because contamination is his worst enemy. No container contamination should ever occur because a subcontractor is unsure about what materials to place in each container. Signage will not make up for poor training and supervision, but it is a necessary supplement to ensure that the training and planning is not defeated by simple confusion.

Not only must the recycling zone containers be clearly signed, but so also should the local hoppers, rolling containers, or vehicles used for collection adjacent to the work area and for transfer to the recycling zone containers.

To reinforce the meaning of the signage, some contractors distribute laminate cards to subcontractor employees with the sign image and bulleted items describing what types of materials correspond with each sign image. While this may seem excessive, the difficulty and cost of correcting a container contamination problem makes it worthwhile to overemphasize the *whats* and *wheres* of the on-site recycling system.

Salvage Materials Checklist

List of materials of interest to architectural salvagers:

- Architectural features
- Banisters
- Bath fixtures
- Bathtubs (mainly white or neutral color)
- Bath vanities
- Bookcases, files, library shelves
- Brick and paving stones
- Cabinetry (wood)
- Claw-foot or antique tubs
- Columns, pillars, and posts
- Concrete blocks and products
- Corbels
- Countertops (straight lengths, neutral colors)
- Displays and display fixtures
- Doors (especially solid wood doors)
- Doors, patio, and French door sets
- Electrical and HVAC supplies
- Faucets and plumbing fixtures
- Fencing
- Flooring (new carpet and vinyl)
- Flooring (wood)
- Glass, sheet, and plexiglass (minimum: 4 ft^2)
- Gutters
- Hinges and other hardware
- Insulation (new or gently used)
- Kitchen cabinet sets
- Kitchen fixtures
- Lighting fixtures
- Lockers
- Lumber (clean, denailed, minimum 4 ft long)
- Mirrors and mirror tiles
- OSB and *Masonite*®
- Plywood and chipboard
- Radiators and registers
- Roof tiles
- Sandstone
- Shelving and racking
- Siding and shutters
- Sinks (kitchen/bath, utility/lab—no chips or other damage)

- Slate, granite, and marble
- Stained glass
- Store fixtures
- Tile (most types and quantities)
- Tile board
- Toilets (low-flow, pre-1940s, no cracks)
- Trim and molding
- Tubs (mainly white or neutral color)
- Windows (wood and vinyl, especially newer energy-efficient windows)
- Wood beams

List of materials not normally of interest to architectural salvagers:

- Appliances older than 5 years
- Ceiling fans
- Commercial 200-V electric equipment
- Commercial bath fixtures
- Commercial ducting and vent covers
- Commercial flashing
- Commercial shelving that are missing parts
- Countertops that are L-shaped or are dated colors
- Doors that are damaged, commercial, or hollow
- Electric baseboard heaters
- Fireplace doors
- Fluorescent light fixtures, bulbs
- Gutters, if leaking or rotted
- Mini-blinds
- Open bags of cement, mortar, or drywall mud
- Room dividers that are missing parts
- Shower doors, except high-end models
- Sinks that are wall hung, cultured marble, or are dated colors
- Tile with heavy grout residue
- Used carpet
- Windows, if aluminum
- Wood lengths shorter than 4 ft, or that are rotten/bug-damaged, contaminated, or nailed
- Wood-burning equipment, unless antique

Density Conversion Factors

SHORT LIST

Volume-to-weight conversion factors for a handful of small, but common, construction products on residential sites.

Note: Some product dimensions are nominal rather than actual. SI conversions are from the nominal dimension.

- *Acoustical tile*: 18 to 23 lb/ft^3 (288 to 368 kg/m^3)
- *Aluminum*: 171 lb/ft^3 (2739 kg/m^3)
- *Asphalt roll roofing*: 70 lb/ft^3 (1121 kg/m^3)
- *Asphalt shingles*: 70 lb/ft^3 (1121 kg/m^3)
- *Blanket and batt insulation, mineral wool, fibrous form*: 0.3 to 2.0 lb/ft^3 (5 to 32 kg/m^3)
- *Boards and slabs insulation*:
  - *Cellular glass*: 8.5 lb/ft^3 (136 kg/m^3)
  - *Expanded polyurethane (R-11 blown)*: 1.5 lb/ft^3 (24 kg/m^3)
  - *Expanded rubber (rigid)*: 4.5 lb/ft^3 (72 kg/m^3)
  - *Glass fiber*: 4 to 9 lb/ft^3 (64 to 144 kg/m^3)
- *Brass:* 524 to 542 lb/ft^3 (8394 to 8682 kg/m^3)
- *Brick, common*: 120 lb/ft^3 (1922 kg/m^3)
- *Brick, face*: 130 lb/ft^3 (2082 kg/m^3)
- *Built-up roofing*: 70 lb/ft^3 (1121 kg/m^3)
- *Cardboard*: 30 lb/yd^3 (18 kg/m^3)
- *Cement mortar*: 116 lb/ft^3 (1858 kg/m^3)
- *Cement plaster, sand aggregate*: 116 lb/ft^3 (1858 kg/m^3)
- *Copper*: 550 to 555 lb/ft^3 (8810 to 8890 kg/m^3)
- *Drywall:* 400 lb/yd^3 (237 kg/m^3)
- *Engineered wood*: 280 lb/yd^3 (166 kg/m^3)
- *Granite, marble*: 150 to 175 lb/ft^3 (2403 to 2803 kg/m^3)
- *Gypsum or plaster board*: 50 lb/ft^3 (801 kg/m^3)
- *Gypsum plaster*: 45 lb/ft^3 (721 kg/m^3)
- *Iron, gray cast*: 438 to 445 lb/ft^3 (7016 to 7128 kg/m^3)
- *Iron, pure*: 474 to 493 lb/ft^3 (7593 to 7897 kg/m^3)
- Lead: 704 lb/ft^3 (11,277 kg/m^3)
- *Masonry*: 1000 lb/yd^3 (593 kg/m^3)
- *Metals*: 150 lb/yd^3 (89 kg/m^3)
- *Mixed wastes*: 95 lb/yd^3 (56 kg/m^3)
- *One-ply membrane roofing*: 83 lb/ft^3 (1330 kg/m^3)
- *Paints, caulks*: 167 lb/yd^3 (99 kg/m^3)
- *Particleboard*: 40 lb/ft^3 (641 kg/m^3)
- *Plywood*: 34 lb/ft^3 (545 kg/m^3)
- *Sand and gravel or stone aggregate*: 140 lb/ft^3 (2243 kg/m^3)
- *Sheathing, fiberboard*: 18 to 25 lb/ft^3 (288 to 400 kg/m^3)
- *Solid sawn wood*: 267 lb/yd^3 (158 kg/m^3)
- *Steel*: 490 lb/ft^3 (7849 kg/m^3)
- *Stucco*: 116 lb/ft^3 (1858 kg/m^3)
- *Vinyl (PVC)*: 150 lb/yd^3 (89 kg/m^3)

LONG LIST

Volume-to-weight conversion factors for a range of bulk commercial demolition materials (see Table 12.1):

- *Ashes, dry*: 1 ft^3 = 35 to 40 lb (1 m^3 = 561 to 641 kg)
- *Ashes, wet*: 1 ft^3 = 45 to 50 lb (1 m^3 = 721 to 801 kg)
- *Asphalt, crushed*: 1 ft^3 = 45 lb (1 m^3 = 721 kg)
- *Asphalt/paving, crushed*: 1 yd^3 = 1380 lb (1 m^3 = 819 kg)
- *Asphalt/shingles comp, loose*: 1 yd^3 = 418.5 lb (1 m^3 = 248 kg)
- *Asphalt/tar roofing*: 1 yd^3 = 2919 lb (1 m^3 = 1732 kg)
- *Bone meal, raw*: 1 ft^3 = 54.9 lb (1 m^3 = 879 kg)
- *Brick, common hard*: 1 ft^3 = 112 to 125 lb (1 m^3 = 1794 to 2002 kg)
- *Brick, whole*: 1 yd^3 = 3024 lb (1 m^3 = 1794 kg)
- *Cement, bulk*: 1 ft^3 = 100 lb (1 m^3 = 1602 kg)
- *Cement, mortar*: 1 ft^3 = 145 lb (1 m^3 = 2323 kg)
- *Ceramic tile, loose 6 in × 6 in (15.24 cm × 15.24 cm)*: 1 yd^3 = 1214 lb (1 m^3 = 720 kg)
- *Chalk, lumpy*: 1 ft^3 = 75 to 85 lb (1 m^3 = 1201 to 1362 kg)
- *Charcoal*: 1 ft^3 = 15 to 30 lb (1 m^3 = 240 to 481 kg)
- *Clay, kaolin*: 1 ft^3 = 22 to 33 lb (1 m^3 = 352 to 529 kg)
- *Clay, potter's dry*: 1 ft^3 = 119 lb (1 m^3 = 1906 kg)
- *Concrete, cinder*: 1 ft^3 = 90 to 110 lb (1 m^3 = 1442 to 1762 kg)
- *Concrete, scrap, loose*: 1 yd^3 = 1855 lb (1 m^3 = 1101 kg)
- *Cork, dry*: 1 ft^3 = 15 lb (1 m^3 = 240 kg)
- *Earth, common, dry*: 1 ft^3 = 70 to 80 lb (1 m^3 = 1121 to 1281 kg)
- *Earth, loose*: 1 ft^3 = 76 lb (1 m^3 = 1217 kg)
- *Earth, moist, loose*: 1 ft^3 = 78 lb (1 m^3 = 1249 kg)
- *Earth, mud*: 1 ft^3 = 104 to 112 lb (1 m^3 = 1666 to 1794 kg)
- *Earth, wet, containing clay*: 1 ft^3 = 100 to 110 lb (1 m^3 = 1602 to 1762 kg)
- *Fiberglass insulation, loose*: 1 yd^3 = 17 lb (1 m^3 = 10 kg)
- *Fines, loose*: 1 yd^3 = 2700 lb (1 m^3 = 1602 kg)
- *Glass, broken*: 1 ft^3 = 80 to 100 lb (1 m^3 = 1281 to 1602 kg)
- *Glass, plate*: 1 ft^3 = 172 lb (1 m^3 = 2755 kg)
- *Glass, window*: 1 ft^3 = 157 lb (1 m^3 = 2515 kg)
- *Granite, broken or crushed*: 1 ft^3 = 95 to 100 lb (1 m^3 = 1522 to 1602 kg)
- *Granite, solid*: 1 ft^3 = 130 to 166 lb (1 m^3 = 2082 to 2659 kg)
- *Gravel, dry*: 1 ft^3 = 100 lb (1 m^3 = 1602 kg)
- *Gravel, loose*: 1 yd^3 = 2565 lb (1 m^3 = 1522 kg)
- *Gravel, wet*: 1 ft^3 = 100 to 120 lb (1 m^3 = 1602 to 1922 kg)
- *Gypsum, pulverized*: 1 ft^3 = 60 to 80 lb (1 m^3 = 962 to 1281 kg)
- *Gypsum, solid*: 1 ft^3 = 142 lb (1 m^3 = 2275 kg)
- *Lime, hydrated*: 1 ft^3 = 30 lb (1 m^3 = 481 kg)
- *Limestone, crushed*: 1 ft^3 = 85 to 90 lb (1 m^3 = 1362 to 1442 kg)
- *Limestone, finely ground*: 1 ft^3 = 100 lb (1 m^3 = 1599 kg)
- *Limestone, solid*: 1 ft^3 = 165 lb (1 m^3 = 2643 kg)

TABLE 12.1   VOLUME AND WEIGHT OF CONCRETE FOR COMMON TYPES OF BUILDING FLOOR SYSTEMS

TYPE OF FLOOR CONSTRUCTION	SPAN	DEAD LOAD + LIVE LOAD		VOLUME AND WEIGHT *PER SQUARE FOOT OF FLOOR AREA*		VOLUME AND WEIGHT *PER SQUARE FOOT OF FLOOR AREA*	
				CONCRETE	REINFORCING	CONCRETE	REINFORCING
ONE-WAY BEAM AND SLAB	15 ft or 4.6 m	100 lb/ft^2	488 kg/m^2	0.42 ft^3	1.90 lb	0.012 m^3	0.86 kg
	20 ft or 6.1 m			0.54 ft^3	2.69 lb	0.015 m^3	1.22 kg
	25 ft or 7.6 m			0.69 ft^3	3.93 lb	0.020 m^3	1.78 kg
TWO-WAY BEAM AND SLAB	15 ft or 4.6 m	100 lb/ft^2	488 kg/m^2	0.47 ft^3	2.26 lb	0.013 m^3	1.03 kg
	20 ft or 6.1 m			0.63 ft^3	3.06 lb	0.018 m^3	1.39 kg
	25 ft or 7.6 m			0.83 ft^3	3.79 lb	0.024 m^3	1.72 kg
FLAT PLATE	15 ft or 4.6 m	100 lb/ft^2	488 kg/m^2	0.46 ft^3	2.14 lb	0.013 m^3	0.97 kg
	20 ft or 6.1 m			0.71 ft^3	2.72 lb	0.020 m^3	1.23 kg
	25 ft or 7.6 m			0.83 ft^3	3.47 lb	0.024 m^3	1.57 kg
FLATE PLATE WITH DROP PANELS	20 ft or 6.1 m	100 lb/ft^2	488 kg/m^2	0.64 ft^3	2.83 lb	0.018 m^3	1.28 kg
	25 ft or 7.6 m			0.79 ft^3	3.88 lb	0.022 m^3	1.76 kg
	30 ft or 9.1 m			0.96 ft^3	4.66 lb	0.027 m^3	2.11 kg
WAFFLE DOMES (30 IN)	25 ft or 7.6 m	50 lb/ft^2	244 kg/m^2	0.69 ft^3	1.83 lb	0.020 m^3	0.83 kg
	30 ft or 9.1 m			0.74 ft^3	2.39 lb	0.021 m^3	1.08 kg
	35 ft or 10.7 m			0.86 ft^3	2.71 lb	0.024 m^3	1.23 kg
	40 ft or 12.2 m			0.98 ft^3	4.80 lb	0.028 m^3	2.18 kg

- *Mortar, hardened*: 1 ft^3 = 100 lb (1 m^3 = 1602 kg)
- *Mortar, wet*: 1 ft^3 = 150 lb (1 m^3 = 2403 kg)
- *Mud, dry close*: 1 ft^3 = 110 lb (1 m^3 = 1762 kg)
- *Mud, wet fluid*: 1 ft^3 = 120 lb (1 m^3 = 1922 kg)
- *Pebbles*: 1 ft^3 = 90 to 100 lb (1 m^3 = 1442 to 1602 kg)
- *Pumice, ground*: 1 ft^3 = 40 to 45 lb (1 m^3 = 641 to 721 kg)
- *Pumice, stone*: 1 ft^3 = 39 lb (1 m^3 = 625 kg)
- *Quartz, sand*: 1 ft^3 = 70 to 80 lb (1 m^3 = 1121 to 1281 kg)
- *Quartz, solid,* 1 ft^3 = 165 lb (1 m^3 = 2643 kg)
- *Rock, loose*: 1 yd^3 = 2570 lb (1 m^3 = 1525 kg)
- *Rock, soft:* 1 ft^3 = 100 to 110 lb (1 m^3 = 1602 to 1762 kg)
- *Sand, dry*: 1 ft^3 = 90 to 110 lb (1 m^3 = 1442 to 1762 kg)
- *Sand, loose*: 1 yd^3 = 2441 lb (1 m^3 = 1448 kg)
- *Sand, moist*: 1 ft^3 = 100 to 110 lb (1 m^3 = 1602 to 1762 kg)
- *Sand, wet*: 1 ft^3 = 110 to 130 lb (1 m^3 = 1762 to 2082 kg)
- *Sewage, sludge*: 1 ft^3 = 40 to 50 lb (1 m^3 = 641 to 801 kg)
- *Sewage, dried sludge*: 1 ft^3 = 35 lb (1 m^3 = 561 kg)
- *Sheetrock scrap, loose*: 1 yd^3 = 394 lb (1 m^3 = 233 kg)
- *Slag, crushed*: 1 yd^3 = 1998 lb (1 m^3 = 1185 kg)
- *Slag, loose*: 1 yd^3 = 2970 lb (1 m^3 = 1762 kg)
- *Slag, solid:* 1 ft^3 = 160 to 180 lb (1 m^3 = 2563 to 2883 kg)
- *Slate, fine ground*: 1 ft^3 = 80 to 90 lb (1 m^3 = 1281 to 1442 kg)
- *Slate, granulated*: 1 ft^3 = 95 lb (1 m^3 = 1522 kg)
- *Slate, solid*: 1 ft^3 = 165 to 175 lb (1 m^3 = 2643 to 2803 kg)
- *Sludge, raw sewage*: 1 ft^3 = 64 lb (1 m^3 = 1025 kg)
- *Soap, chips*: 1 ft^3 = 15 to 25 lb (1 m^3 = 249 to 400 kg)
- *Soap, powder*: 1 ft^3 = 20 to 25 lb (1 m^3 = 320 to 400 kg)
- *Soap, solid*: 1 ft^3 = 50 lb (1 m^3 = 801 kg)
- *Soil/sandy loam, loose*: 1 yd^3 = 2392 lb (1 m^3 = 1419 kg)
- *Stone or gravel*: 1 ft^3 = 95 to 100 lb (1 m^3 = 1522 to 1602 kg)
- *Stone, crushed*: 1 ft^3 = 100 lb (1 m^3 = 1602 kg)
- *Stone, crushed, size reduced*: 1 yd^3 = 2700 lb (1 m^3 = 1602 kg)
- *Stone, large*: 1 ft^3 = 100 lb (1 m^3 = 1602 kg)
- *Wax*: 1 ft^3 = 60.5 lb (1 m^3 = 969 kg)
- *Wood ashes*: 1 ft^3 = 48 lb (1 m^3 = 769 kg)

DEMOLITION MATERIAL ESTIMATING GUIDE (SHORT)

Volume to Pounds

- *Acoustical ceiling tiles*: 4 ft^2: 1.05 (4 m^2 = 21 kg)
- *Asphalt shingles*: 1 ft^2: 3 (3 m^3 = 15 kg)
- *Asphalt shingles*: 1 yd^3: 419 (1 m^3 = 249 kg)
- *Carpeting*: 3 ft^2: 0.5 (3 m^2 = 7 kg)

- *Demolition*: 1 ft^2: 40 (1 m^2 of wall area = 195 kg)
- *Demolition*: 1 lineal foot of wall area: 40 (1 lineal meter of wall area = 2381 kg)
- *Dirt, brick, concrete, and asphalt*: 1 yd^3: 2000 (1 m^3 = 1187 kg)
- *Green waste*: 1 yd^3: 500 (1 m^3 = 297 kg)
- *Gypsum board*: 1 yd^3: 394 (1 m^3 = 234 kg)
- *Metals*: 1 yd^3: 906 (1 m^3 = 538 kg)
- *Mixed C & D materials*: 1 yd^3: 500 (1 m^3 = 297 kg)
- *Renovation debris (general)*: 2 ft^2: 10 (2 m^2 = 49 kg)
- *Wood*: 1 yd^3: 375 (1 m^3 = 222 kg)
- *Wood shake shingle roofing*: 1 ft^2: 2 (1 square meter = 10 kg)
- *Wood shake shingle roofing*: 1 yd^3: 435 (1 m^3 = 258 kg)

DEMOLITION MATERIAL ESTIMATING GUIDE (DETAILED)

Site Materials (lb/ft^3)

- *Cinders or ashes*: 40 to 45 (641 to 721 kg/m^3)
- *Clay, dry*: 63 (1009 kg/m^3)
- *Clay, damp and plastic*: 110 (1762 kg/m^3)
- *Clay and gravel, dry*: 100 (1602 kg/m^3)
- *Earth, dry and loose*: 76 (1217 kg/m^3)
- *Earth, dry and packed*: 95 (1522 kg/m^3)
- *Earth, moist and loose*: 78 (1249 kg/m^3)
- *Earth, moist and packed*: 96 (1538 kg/m^3)
- *Earth, muddy and packed*: 115 (1842 kg/m^3)

Stone (lb/ft^2)

- *Granite*: 165 (806 kg/m^2)
- *Limestone*: 165 (806 kg/m^2)
- *Marble*: 165 (806 kg/m^2)
- *Sandstone/bluestone*: 147 (718 kg/m^2)
- *Slate*: 175 (854 kg/m^2)

Metals (lb/ft^3)
(See Table 12.2.)

- *Brass, red*: 546 (8746 kg/m^3)
- *Brass, yellow*: 528 (8458 kg/m^3)
- *Bronze, commercial*: 552 (8842 kg/m^3)
- *Copper*: 556 (8906 kg/m^3)
- *Iron, cast gray*: 450 (7208 kg/m^3)
- *Iron, wrought*: 485 (7769 kg/m^3)
- *Lead*: 710 (11373 kg/m^3)
- *Monel metal*: 552 (8842 kg/m^3)

TABLE 12.2 AVERAGE WEIGHTS OF STEEL OPEN WEB JOISTS OF VARYING DEPTHS	
DEPTH *IN*	WEIGHT *POUNDS PER LINEAL FOOT*
8	4.2
10	5.1
12	6.2
14	7.6
16	8.9
18	9.8
20	10.2
22	10.8
24	11.5

- *Nickel*: 555 (8890 kg/m^3)
- *Stainless steel*: 500 (8009 kg/m^3)
- *Rolled steel*: 490 (7849 kg/m^3)
- *Zinc*: 440 (7048 kg/m^3)

Concrete (lb/ft^3)

- *Slab, reinforced*: 150 (2403 kg/m^3)
- *Slab, unreinforced*: 144 (2307 kg/m^3)

Mortar & Plaster (lb/ft^3)

- *Mortar, masonry*: 116 (1858 kg/m^3)
- *Plaster, gypsum, sand*: 112 (1794 kg/m^3)
- *Plaster, gypsum, perlite*: 53 (849 kg/m^3)
- *Plaster, Portland cement, sand*: 112 (1794 kg/m^3)
- *Plaster, Portland cement, perlite*: 52 (833 kg/m^3)
- *Plaster, Portland cement, vermiculite*: 52 (833 kg/m^3)

Brick and Block with Mortar (lb/ft^2)

- *4 in (10.2 cm) brick*: 35 (171 kg/m^2)
- *4 in (10.2 cm) concrete block, stone or gravel*: 34 (166 kg/m^2)
- *4 in (10.2 cm) concrete block, lightweight aggregate*: 22 (107 kg/m^2)

- *6 in (15.24 cm) concrete block, stone or gravel:* 50 (244 kg/m^2)
- *6 in (15.24 cm) concrete block, lightweight aggregate:* 31 (151 kg/m^2)
- *8 in (20.3 cm) concrete block, stone or gravel:* 58 (283 kg/m^2)
- *8 in (20.3 cm) concrete block, lightweight aggregate:* 36 (176 kg/m^2)
- *12 in (30.5 cm) concrete block, stone or gravel:* 90 (439 kg/m^2)
- *12 in (30.5 cm) concrete block with lightweight aggregate:* 58 (283 kg/m^2)

Partitions (lb/ft^2)

- *2 in ×4 in (5.0 cm × 10.2 cm) wood stud, lath and plaster:* 15 (73 kg/m^2)
- *4 in (10.2 cm) gypsum block, plaster:* 25 (122 kg/m^2)
- *3 in (7.6 cm) solid plaster:* 27 (132 kg/m^2)
- *4 in (10.2 cm) metal stud, lath and plaster:* 18 (88 kg/m^2)
- *2 in (5.1 cm) solid plaster:* 18 (88 kg/m^2)
- *Moveable steel partitions (office):* 6 (29 kg/m^2)

Glass (lb/ft^2)

- *1/4 in (6.35 mm) thick wire glass:* 3.5 (17 kg/m^2)
- *Double strength 1/8 in (3.17 mm) thickness:* 3 (15 kg/m^2)
- *Glass block, 4 in (10.16 cm) thick:* 20 (98 kg/m^2)
- *Insulated 1/8 in (3.17 mm) thick plate with air space:* 3.25 (16 kg/m^2)
- *Polished plate 1/4 in (6.35 mm) thick:* 3 (15 kg/m^2)
- *Polished plate 1/2 in (12.7 mm) thick:* 7 (34 kg/m^2)

Insulation (lb/ft^2)

- *Fiberglass batts, blankets per 1 in (2.54 cm) thickness:* 0.3 (1 kg/m^2)
- *Rigid board per 1 in (2.54 cm) thickness:* 2.5 (12 kg/m^2)

Acoustical Ceiling (lb/ft^2)

- *Acoustic plaster on gypsum lath:* 11 (54 kg/m^2)
- *Mineral fiber tile ¾ in (19 mm) thick, 12 ×12 in:* 1.4 (7 kg/m^2)
- *Wood fiber board 5/8 in (15.9 mm) thick, 24 ×48 in (61 ×122 cm):* 1 (5 kg/m^2)
- *Wood fiber board 1 in (2.54 cm) thick, 24 ×48 in (61 ×122 cm):* 1.5 (7 kg/m^2)

Stone Veneer (No-Mortar) (lb/ft^2)

- *2 in (5.1 cm) granite, ½ in (12.7 mm) parging:* 30 (146 kg/m^2)
- *4 in (10.2 cm) granite, ½ in (12.7 mm) parging:* 59 (288 kg/m^2)
- *6 in (15.2 cm) limestone facing, ½ in (12.7 mm) parging:* 55 (269 kg/m^2)
- *4 in (10.2 cm) sandstone/bluestone, 1/2 in (12.7 mm) parging:* 49 (239 kg/m^2)
- *1 in (2.54 cm) marble:* 13 (63 kg/m^2)
- *1 in (2.54 cm) slate:* 14 (68 kg/m^2)

Roofing Materials (lb/ft²)

- *Asbestos cement shingles:* 3 (15 kg/m²)
- *Asphalt shingles*: 2 (10 kg/m²)
- *Built-up*: 7 (34 kg/m²)
- *Copper:* 2 (10 kg/m²)
- *Corrugated asbestos*: 4 (20 kg/m²)
- *Deck, steel without roofing or insulation*: 3 (15 kg/m²)
- *Fiberglass panels (2-1/2 in, or 6.35 cm, corrugations)*: 6 (29 kg/m²)
- *Monel metal*: 1 (5 kg/m²)
- *Plastic sandwich panel, 2-1/2 in (or 6.35 cm) thick*: 3 (15 kg/m²)
- *Slate, 1/4 in (6.35 mm) thick*: 8 (39 kg/m²)
- *Slate, 1/2 in (12.7 mm) thick*: 16 (78 kg/m²)
- *Steel*: 3 (15 kg/m²)
- *Tile, cement flat*: 13 (63 kg/m²)
- *Tile, cement ribbed:* 16 (78 kg/m²)
- *Tile, clay flat w/setting bed*: 18 (88 kg/m²)
- *Tile, clay mission*: 14 (68 kg/m²)
- *Tile, clay shingle type*: 12 (59 kg/m²)
- *Wood shingles*: 2 (10 kg/m²)

Finish Materials (lb/ft²)

- *Acoustical tile, 1/2 in (12.7 mm) thick with no grid*: 1 (5 kg/m²)
- *Building board, 1/2 in (12.7 mm) thick*: 8.0 (39 kg/m²)
- *Cement finish, 1 in (2.54 cm) thick:* 12 (59 kg/m²)
- *Gypsum board, 1/2 in (12.7 mm) thick*: 2 (10 kg/m²)
- *Hardwood flooring, 25/32 in (19.8 mm) thick*: 4 (20 kg/m²)
- *Marble w/setting bed*: 28 (137 kg/m²)
- *Plaster, suspended w/lath*: 10 (49 kg/m²)
- *Plywood, 1/2 in (12.7 mm) thick*: 2 (10 kg/m²)
- *Quarry tile, 1/2 in (12.7 mm) thick*: 6 (29 kg/m²)
- *Quarry tile, 3/4 in (19.1 mm) thick*: 9 (44 kg/m²)
- *Slate, 3/16 in to 1/4 in (4.76 mm to 6.35 mm):* 7.0 to 9.5 (34 kg to 46 kg/m²)
- *Terrazzo, 2 in (5.1 cm) thick*: 25 (122 kg/m²)
- *Terrazzo, 3 in (7.6 cm)*: 38.0 (186 kg/m²)
- *Tile, ceramic mosaic 1/4 in (6.35 mm) thick*: 3 (15 kg/m²)
- *Tile, glazed wall 3/8 in (9.5 mm) thick*: 3 (15 kg/m²)
- *Vinyl asbestos tile, 1/2 in (12.7 mm) thick*: 1.33 (6 kg/m²)

Floor and Roof Construction (lb/ft²)

- *Concrete-lightweight, plain 1 in (2.54 cm) thick:* 5 (24 kg/m²)
- *Concrete-lightweight, reinforced 1 in (2.54 cm) thick*: 8 (39 kg/m²)

- *Concrete-slag, plain 1 in (2.54 cm) thick*: 11 (54 kg/m^2)
- *Concrete-slag, reinforced 1 in (2.54 cm) thick*: 11 (54 kg/m^2)
- *Concrete-stone, plain 1 in (2.54 cm) thick*: 12 (59 kg/m^2)
- *Concrete-stone, reinforced 1 in (2.54 cm) thick*: 12 (59 kg/m^2)
- *Flexcore, 6 in (15.24 cm) pre-cast lightweight concrete*: 30 (146 kg/m^2)
- *Flexcore, 6 in (15.24 cm) pre-cast stone concrete*: 40 (195 kg/m^2)
- *Plank, cinder concrete 2 in (5.1 cm) thick*: 15 (73 kg/m^2)
- *Plank, durisol roof 3-1/4 in to 4-1/4 in thick (8.3 cm to 10.8 cm)*: 14 to 17 (68 to 83 kg/m^2)
- *Plank, gypsum 2 in (5.1 cm) thick*: 12 (59 kg/m^2)

A

GOVERNMENT AND INDUSTRY LINKS

These listings were accurate as of the date they were compiled. However—even in the case of governmental agencies—locations, phone numbers, Web site URLs, and e-mail addresses may be subject to change at any time. It is best to re-verify contact information for these entities before sending them vital correspondence, records, or other documentation.

Web Resources

Georgia

Georgia Recycling Coalition (My Ecoville): www.myecoville.com/us/ga/home

National

Environmental Industry Associations: www.envasns.org
Environmental Research and Education Foundation: www.erefdn.org
National Solid Wastes Management Association: www.nswma.org

State and Municipal Government Resources

Alabama

Solid Waste Branch
Alabama Department of Environmental Management
P.O. Box 301463
Montgomery, AL 36130-1463
334.271.7771, fax 334.279.3050

State Recycling Coordinator
Alabama Department of Economic and Community Affairs
P.O. Box 5690
Montgomery, AL 36103-5690
334.242.5322

Alaska

Environmental Specialist
Alaska Department of Environmental Conservation
410 Willoughby Avenue, Suite 303
Juneau, AK 99801
907.465.5153, fax 907.465.5362

Recycling Public Assistance
Alaska Department of Environmental Conservation
555 Cordova Street
Anchorage, AK 99567
907.269.7582

Arizona

Recycling Coordinator
Arizona Department of Environmental Quality
1110 West Washington
Phoenix, AZ 85007
602.771.4134, fax 602.771.2382

Arkansas

Solid Waste Division
Arkansas Department of Environmental Quality
P.O. Box 8913
Little Rock, AR 72219-8913
501.682.0814

California

Alameda County Waste Management Authority
Recycling education and grant programs: www.stopwaste.org/
County of San Diego, CA
Unincorporated County Recycling and Household Hazardous Waste Hotline
877.R.1.EARTH

California Integrated Waste Management Board
8800 Cal Center Drive
Sacramento, CA 95826
916.255.2425, fax 916.255.4207

Closed, Illegal, and Abandoned Site Investigation Section
Permitting and Enforcement
California Integrated Waste Management Board
1001 I Street, Mail Stop 20
Sacramento, CA 95812
916.341.6723, fax 916.319.7552

Colorado

Colorado Department of Public Health and Environment
Hazardous Materials and Waste Management Division HMWMD-B2
4300 Cherry Creek Drive South
Denver, CO 80246-1530
303.692.3300

Connecticut

Recycling Unit
Waste Planning and Standards Division
Connecticut Department of Environmental Protection
79 Elm Street, 4th Floor
Hartford, CT 06106-5127
860.424.3130, fax 860.424.4081

A printed list of permitted C&D processing facilities is available from the Connecticut Department of Environmental Protection at 860.424.3365. A listing of aggregate recycling facilities is available at www.dep.state.ct.us/wst/recycle/construct.

Delaware

Solid and Hazardous Waste Management Branch
Delaware Department of Natural Resources and Environmental Control
89 Kings Highway
302.739.3689, fax 302.739.5060

Florida

Solid Waste Section
Florida Department of Environmental Protection
2600 Blair Stone Road, MS 4565
Tallahassee, FL 32399
850.245.8734, fax 850.245.8811

Georgia

Environmental Protection Division
Georgia Environmental Protection Division
4244 International Parkway, Suite 104
Atlanta, GA 30354
404.362.4510, fax 404.362.2693

Hawaii

Solid and Hazardous Waste Branch
Hawaii Department of Health
919 Ala Moana Boulevard, #212
Honolulu, HI 96814
808.586.4245, fax 808.586.7509

Idaho

Idaho Department of Environmental Quality
1410 North Hilton
Boise, ID 83706
208.373.0416

Illinois

Division of Land Pollution Control
Illinois Environmental Protection Agency
1021 North Grand Avenue East
Springfield, IL 62702
217.524.3306, fax 217.524.3291

Indiana

Solid Waste Technical Compliance Section
Indiana Department of Environmental Management
100 North State Avenue
P.O. Box 6015
Indianapolis, IN 46206-6015
317.308.3110, 800.451.6027, fax 317.232.3403

Iowa

Energy and Waste Management Bureau
Iowa Department of Natural Resources
502 East Ninth Street
Des Moines, IA 50319-0034
515.281.8176, fax 515.281.8895

Kansas

Chief, Markets and Development
Bureau of Waste Management
Kansas Department of Health and Environment
1000 SW Jackson, Suite 320
Topeka, KS 66612-1366
785.296.1600, fax 785.296.8909
www.kdhe.state.ks.us/waste

Kentucky

Division of Waste Management
Kentucky Department for Environmental Protection
14 Reilly Road
Frankfort, KY 40601
502.564.6716, fax 502.564.4049

Louisiana

Solid Waste Division
Louisiana Department of Environmental Quality
P.O. Box 82135
Baton Rouge, LA 70884-2178
225.765.0355, fax 225.765.0617

Maine

Division of Solid Waste Facilities Regulation
Maine Department of Environmental Protection
State House: Station 17
Augusta, ME 04333-0017
207.287.7718, fax 207.287.7826

Environmental Specialist
Maine Department of Environmental Protection
312 Canco Road
Portland, ME 04103
207.822.6343, fax 207.822.6303

Haulers: Contact municipal recycling officials in the community in which a project is located (directory of recycling officials.

Markets: The Maine State Planning Office offers a searchable Waste Management Services Directory at www.maine.gov/spo/recycle/

Maryland

Chief Recycling Division
Maryland Department of Environment
1800 Washington Boulevard, Suite 610
Baltimore, MD 21230-1719
410.537.3314

Massachusetts

Massachusetts Department of Environmental Protection
Bureau of Waste Prevention
Business Compliance Division, 9th Floor
One Winter Street
Boston, MA 02108
617.574.6867, fax 617.292.5778

Haulers: A listing of haulers is available at: www.mass.gov/dep/recycle

Markets: The Recycling Services Directory maintained by Massachusetts WasteCap is available at: www.wastecap.org

Michigan

Department of Environmental Quality
Waste and Hazardous Material Division
P.O. Box 30241
Lansing, MI 48909
517.335.4712, fax 517.373.4797

Minnesota

Policy and Planning Division
Minnesota Pollution Control Agency
520 Lafayette Road
St. Paul, MN 55155
651.296.8745, fax 651.297.8676

Mississippi

Solid Waste Branch
Mississippi Department of Environmental Quality
P.O. Box 10385
Jackson, MS 39289-0385
601.961.5304, fax 601.354.6612

Missouri

Engineering Section
Missouri Department of Natural Resources
P.O. Box 176
Jefferson City, MO 65102
573.751.5401, fax 573.526.3902

Montana

Planning, Prevention, and Assistance Division
Lee Metcalf Building (Main Office)
1520 East Sixth Avenue
P.O. Box 200901
Helena, MT 59620-0901

Nebraska

Waste Management Section
Nebraska Department of Environmental Quality
P.O. Box 98922
Lincoln, NE 68509-8922
402.471.8308, fax 402.471.2909

Nevada

Solid Waste Branch Supervisor
Nevada Division of Environmental Protection
333 West Nye Lane
Carson City, NV 89706
775.687.9468, fax 775.687.6396

New Hampshire

Waste Management
NH Department of Environmental Services
6 Hazen Drive
P.O. Box 95
Concord, NH 03302-0095
603.271.1373, fax 603.271.1381

Haulers: Haulers who have identified themselves to the New Hampshire Department of Environmental Services are listed at http://des.nh.gov/.

Markets: The New Hampshire Department of Environmental Services, Solid Waste Technical Assistance Section provides market information at http://des.nh.gov/.

New Jersey

Bureau of Resource Recovery and Technical Programs
Division of Solid and Hazardous Waste
New Jersey Department of Environmental Protection
P.O. Box 414
Trenton, NJ 08625-0414
609.984.6985, fax 609.633.9839

New Mexico

Program Manager
Solid Waste Bureau
Harold South Runnels Building
1190 St. Francis Drive
Santa Fe, NM 87502-0110
505.827.2952

New York

Beneficial Use Section
New York State Department of Environmental Conservation
625 Broadway, 9th Floor
Albany, NY 12233-7253
518.402.8706, fax 518.402.9024

New York State Department of Environmental Conservation
625 Broadway, 9th Floor
Albany, NY 12233-7253
518.402.8706, fax 518.402.9024

North Carolina

Division of Waste Management
North Carolina Department of Environment and Natural Resources
401 Oberlin Road
Suite 150
Raleigh, NC 27605
919.733.0692 x 260, fax 919.715.3605

North Dakota

North Dakota Department of Health
1200 Missouri Avenue
P.O. Box 5520
Bismarck, ND 58506-5520
701.328.5166

Ohio

Division of Solid and Infectious Waste Management
Ohio Environmental Protection Agency
P.O. Box 1049
Columbus, OH 43216-1049
614.728.5373, fax 614.728.5315

Oklahoma

Waste Management Division
Oklahoma Department of Environmental Quality
P.O. Box 1677
Oklahoma City, OK 73101-1677
405.702.5100, fax 405.702.5151

Oregon

Oregon Division of Environmental Quality
2020 SW Fourth Avenue, Suite 400
Portland, OR 97201
503.229.5364

Pennsylvania

Division of Municipal and Residual Waste
Pennsylvania Department of Environmental Protection
Rachel Carson State Office Building, 14th Floor
P.O. Box 8472
Harrisburg, PA 17105-8472
717.787.7381, fax 717.787.1904

Rhode Island

Office of Waste Management
Rhode Island Department of Environmental Management
235 Promenade Street
Providence, RI 02908
401.222.2797

Markets: A printed list of the licensed C&D processing facilities in Rhode Island is available from the Rhode Island Department of Environmental Management at 401.222.2797.

South Carolina

Solid Waste Planning and Recycling
South Carolina Department of Health and Environmental Control
2600 Bull Street
Columbia, SC 29201
803.896.4202, fax 803.896.4001

South Dakota

Waste Management Program
Division of Environmental Services
South Dakota Department of Environment and Natural Resources
523 East Capitol
Pierre, SD 57501-3182
605.773.3153, fax 605.773.6035

Tennessee

Solid Waste Management Unit
Tennessee Department of Environment and Conservation
401 Church Street, 5th Floor
Nashville, TN 37243-1535
615.532.0818, fax 615.532.0886

Texas

Regulation and Evaluation Division
Texas Natural Resources Conservation Commission
P.O. Box 13087 (MC 129)
Austin, TX 78711-3087
512.239.6412, fax 512.239.6410

Utah

Division of Solid and Hazardous Waste
P.O. Box 144880
Salt Lake City, UT 84114-4880
801.538.6794, fax 801.538.6715

Vermont

Vermont Department of Environmental Conservation
Solid Waste Management Program
103 South Main Street
South Waterbury, VT 05671-0407
802.241.3481, fax 802.244.5141

Haulers: A listing of Vermont haulers is posted at: www.anr.state.vt.us/dec/wastediv/rcra/pubs/AllTrans.pdf.

Markets: The Vermont Agency of Natural Resources maintains a Web site with some market information: www.anr.state.vt.us.

Virginia

Virginia Department of Environmental Quality
P.O. Box 10009
Richmond, VA 23240
804.698.4029

Washington

Washington State Department of Ecology
Solid Waste and Financial Assistance Program
P.O. Box 47775
Olympia, WA 98504-7774
360.407.6383, fax 360.407.6305

Seattle/King County, WA

201 South Jackson Street
Suite 701
Seattle, WA 98104-3855
800.325.6165, 206.296.4434
www.kingcounty.gov

C&D Recycling Program
Link-up Program
Recycling Directory
Solid Waste Division

West Virginia

Solid Waste Permitting Unit
West Virginia Department of Environmental Protection
1356 Hansford Street
Charleston, WV 25301
304.558.6350, fax 304.558.1574

West Virginia Department of Natural Resources
Recycling Coordinator
Capitol Complex Building 3, Room 732
1900 Kanawha Boulevard East
Charleston, WV 25305
304.558.3370

Wisconsin

Waste Management—Air and Waste Division
Wisconsin Department of Natural Resources
1125 North Military Avenue, Box 10448
Green Bay, WI 54307-0448
920.492.5867, fax 920.492.5867

Wisconsin Department of Natural Resources
3911 Fish Hatchery Road
Fitchburg, WI 53711
608.275.3466, fax 608.275.3338

Wyoming

Solid Waste Permitting and Corrective Action
Solid and Hazardous Waste Division
Wyoming Department of Environmental Quality
3030 Energy Lane, Suite 200
Casper, WY 82604
307.473.3450, fax 307.473.3458

General Resources

Associated General Contractors (AGC) (http://www.agc.org/)

■ Supports recycling by its members. The association produced a brochure that provides examples of recycling projects carried out by AGC members, along with facts and statistics pertaining to the reclamation of asphalt, concrete, steel, and wood.

California Resource Recovery Association Construction and Demolition Council (www.crra.com/cdc/index.html)

■ Promotes the advancement of C&D material recovery practices and strives to expand C&D recovery infrastructure and markets.

CalRecycle (www.calrecycle.ca.gov/)

■ CalRecycle merges the duties of the California Integrated Waste Management Board and the California DOC's Division of Recycling to best protect public health and the environment by efficiently managing California's waste disposal and recycling efforts.

Construction Materials Recycling Association (www.cdrecycling.org)

■ Provides information on issues and technology facing the industry, including a listing of available literature on relevant topics; promotes the acceptance and use of recycled construction materials.

Inform (www.informinc.org)

■ An independent research organization with a focus on waste prevention, including strategies for reducing waste and preventing pollution generated during building construction, renovation, and demolition.

Maryland Recycles (http://www.mdrecycles.org/index.asp)

■ Maryland's official recycling site for government, business, and homeowners.

Mid-Atlantic Consortium of Recycling and Economic Development Officials (MACREDO) (www.libertynet.org/macredo/)

■ A six-member regional organization that seeks to identify, promote, and implement projects and programs that enhance recycling and economic development.

National Association of Demolition Contractors (NADC) (www.demolitionassociation.com)

■ Represents contractors that manage demolition debris, including disposal and recycling.

National Demolition Association (www.demolitionassociation.com/): The National Demolition Association represents more than 1000 U.S. and Canadian companies that offer standard demolition services as well as a full range of demolition-related services.

Northeast Recycling Council (www.nerc.org)

■ Recycling news, resources, and links for 10 Northeastern states.

Northeast Resource Recovery Association (www.recyclewithus.org/)

■ New England's leading recycling agency.

Reuse Development Organization (ReDO) (www.redo.org)

■ ReDO is a clearinghouse for information on reuse opportunities nationwide. ReDO helps companies redistribute donations of materials too big for a single reuse center. Their Web site contains a good discussion of the environmental, community, and economic benefits of reuse.

Seattle/King County Business and Industry Venture (Washington State) (www. resourceventure.org/construction.htm)

■ Construction recycling guide for contractors.

University of Florida Center for Construction and the Environment (www.cce.ufl.edu)

■ Specializes in research on sustainable construction, construction and demolition waste recycling, and deconstruction.

U.S. Environmental Protection Agency (www.epa.gov)

■ Construction and demolition debris Web site including deconstruction information and case studies.

U.S. Environmental Protection Agency (www.epa.gov)

■ Section of the Environmental Protection Agency Web site that deals with construction waste issues, including links to other resources.

U.S. Green Building Council (www.usgbc.org)

■ Leading coalition of the green building industry, and originators of the LEED rating system.

U.S. Occupational Safety and Health Administration (OSHA) (www.osha.gov)

■ OSHA regulations place restrictions on the manual handling of asbestos and lead-based paint. Visit OSHA online and follow the link to "advisors" to download programs to assist in determining responsibilities.

WasteSpec (www.tjcog.dst.nc.us/cdwaste.htm)

■ Provides free detailed model specification language for reducing and recycling construction and demolition debris.

National Recyclers

Note: Most recycling operations are local to the areas they serve. This guide lists companies that either provide nationwide resources or host databases that list local recycling companies.

General Recycling

Construction Waste Management Database (www.wbdg.org)

■ A searchable database of C&D recyclers across the United States, created by the General Services Administration, National Institute of Standards and Technology, the U.S. Environmental Protection Agency, and other partners.

Habitat for Humanity (www.habitat.org)

■ National low-income housing provider. Hundreds of local chapters reuse architectural salvage and other demolition materials.

Loading Dock (www.loadingdock.org)

■ This nonprofit building materials reuse organization is based in Baltimore, Maryland.

RecycleNet (www.recycle.net)

■ An extensive online listing of markets for all types of recycled products. Includes links to trade and industry associations, as well as current market prices for high-value items.

Building Materials Reuse Association (www.bmra.org)

■ A nonprofit, membership-based organization representing companies and organizations involved in the acquisition and/or redistribution of used building materials.

Waste Management (866.WMCYCLE or www.wm.com)

■ Waste Management, Inc. is the leading provider of comprehensive waste and environmental services in North America.

Weyerhauser Corporation Recycling (www.weyerhaeuser.com. Also: http://www.weyerhaeuser.com/)

■ Weyerhauser runs active paper and plastic recycling activities.

Online Materials Exchange

Georgia Industrial Materials Exchange: www.ScrapMatchGA.org
Industrial Materials Exchange (IMEX): www.govlink.org/hazwaste/business/imex
New York WasteMatch (212.442.5219, www.wastematch.org)
North Carolina Materials Exchange: www.ncwastetrader.org
South Carolina Materials Exchange: www.scdhec.gov/environment
Southeast Waste Exchange: 704.547.2270
Southern Waste Information Exchange: www.wastexchange.org
The Northeast Industrial Waste Exchange: 315.422.6572

■ New York City materials exchange service, linking generators of industrial and commercial waste by-products with local end-users.

Carpet

Antron Reclamation Center: Calhoun, GA. 706.624.8833
Atlanta Foam Recycle: Atlanta, GA. 713.201.0068
Carpet America Recovery Effort (CARE): Nationwide carpet recycling program.
    www.carpetrecovery.org
Columbia Recycling: Dalton, GA. 706.278.4701
Evergreen Nylon (Shaw): Augusta, GA. 800.434.9887
Foam Recycle Center: Savannah, GA. 912.667.7992
Interface Americas: LaGrange, GA. 706.812.6193
Invista, Inc.: Calhoun, GA. Facility accepts used carpet from commercial sources and installers. 770.420.7809, http://antron.net/
Milliken Carpet: Nationwide carpet recycling program. 888.645.6239,
    www.milliken.com

Hazardous and Universal Waste

National Electrical Manufacturers Association (NEMA) (www.nema.org/lamprecycle)

■ The site provides a list of association companies that process or recycle spent lamps containing mercury.

Per Scholas (www.perscholas.org)

■ Organization that recycles computers for use in low-income schools in Bronx, NY.

Bethlehem Apparatus Company (610.838.7034, www.bethapp.com, mail@bethapp.com)

■ Recycles and distills mercury waste at

890 Front Street
Hellertown, PA 18055

Expanded Polystyrene (EPS)

Alliance of Foam Packaging Recyclers
1298 Cronson Boulevard, Suite 201
Crofton, MD 21114
410.451.8340, fax 410.451.8343

Plastic Loose Fill Council (PLFC)

■ Contact them at www.loosefillpackaging.com/pages/why.htm, or call the Peanut Hotline, 800.828.2214, a toll-free, automated 24-hour service that provides callers with the location of the nearest site that accepts loose fill packaging for reuse.

Gypsum Board (drywall)

Waste Grinding: http://wastegrinding.com/

Ceiling Tiles

Armstrong World Industries: 877.276.7876, www.armstrong.com/commceilingsna/environmental.html

Debris and Miscellaneous

National Wooden Pallet and Container Association: 703.519.6104, www.nwpca.com, ecoppage@palletcentral.com

Roof Shingles

Shingle Recycling: www.shinglerecycling.org

Wood Waste

www.woodfuel.com
Wood Planet: www.woodplanet.com

American Forest and Paper Association (AFPA)

■ Publishes the *National Wood Recycling Directory*, a listing of recycling sites that accept wood waste, including construction debris. It costs $5. To order a copy, call AFPA at 202.462.2700.

B

GLOSSARY

Abatement debris: Waste from hazardous material remediation activities.

Aggregate: Material used in paving, consisting of hard, graduated fragments of inert mineral materials, including sand, gravel, crushed stone, slag, rock dust, or powder.

Appropriate for processing: Loads of C&D materials entering a facility, of which most of the material (approximately 90 percent), as determined by the processor, can be sent on by the facility for recycling. Such loads are usually charged a lower rate by the recycler.

Architectural salvage: Existing building materials or features removed by a salvage contractor before or during the demolition process. Architectural salvage value is considered to be waste that is diverted from landfills.

Asbestos: An incombustible, chemical-resistant, fibrous mineral form of impure magnesium silicate. Prolonged inhalation of asbestos can lead to asbestosis— a chronic, progressive lung disease. Asbestos is often found in friable and non-friable forms in plaster, floor tiles, pipe and boiler insulation, and fire-protection products.

Asbestos abatement: Procedures to control fiber release from asbestos-containing materials in a building or to remove them entirely, including removal, encapsulation, repair, enclosure, encasement, and operations and maintenance programs.

Bale: A large block of recyclables held together with plastic strapping.

Baler: A machine that compacts waste materials, usually into rectangular bales. Balers often are used on newspaper, plastics, and corrugated cardboard. Balers compress saleable ferrous scrap into a more uniform, rectangular shape (sometimes referred to as a "log") and enable scrap metal recyclers to move baled material more efficiently from the yard to the shredder or to the mill.

Brown goods: Bulky residential items that are difficult to recycle, such as mattresses and furniture.

Building permit: Approval issued by the appropriate local building or construction department allowing construction of the project to proceed. Building permits are often issued for various components of the work, including: demolition; foundation/footing; fire protection; electrical; and plumbing. Also known as a construction permit.

By-product: A secondary product of a manufacturing process. A waste by-product is an unwanted by-product that can either be disposed of or recycled.

Ceiling tiles: These panels, also called acoustical ceiling tiles, are made from a variety of materials and are designed to reduce noise. Many recyclers will accept mineral fiber-based tiles that are free of contaminants but will not normally accept cast tile, fiberglass board, ceramic-based, or laminated tiles.

Certified asbestos contractor: A contractor licensed by the state or municipality to conduct asbestos-containing materials removal and disposal.

Clean: Untreated and unpainted material; not contaminated with oils, solvents, caulk, or other contaminants that would render the material unusable by a recycler.

Clean rubble: Inert uncontaminated construction and demolition waste, which can include asphalt, concrete and concrete products, reinforcing steel, brick and concrete masonry units, and soil or rock.

Commercial waste: Waste material that originates in wholesale business establishments, office buildings, stores, schools, hospitals, and government agencies. Also known as retail waste.

Commingled recycling: A type of recycling that allows contractors to put types of waste into common containers to save space and labor. The commingled materials, typically consisting of wood, metal, and cardboard, are separated at the recycling facility for processing. Commingled recycling is less expensive than landfill disposal but typically more expensive than source-separated recycling. This type of recycling also makes it more difficult to accurately track material recycling and disposal by material type.

Compost: Decomposed organic material resulting from the composting process. Used to enrich or improve the consistency of soil.

Composting: A waste management technique involving the controlled biological decomposition of organic materials into a stable product that can be applied to the land without adversely affecting the environment.

Construction and demolition landfill: A permitted and approved solid waste disposal area used solely for the disposal of construction and demolition wastes. This does not include a site that is used exclusively for the disposal of clean rubble.

Construction and demolition waste (C&D waste): Solid waste resulting from the construction, remodeling, repair, and demolition of structures, roads, sidewalks, and utilities. Such wastes include, but are not limited to: bricks, concrete and other masonry materials; roofing materials; soil; rock; wood and/or wood products;

wall or floor coverings; plaster; drywall; plumbing fixtures; electrical wiring; electrical components containing no hazardous materials; non-asbestos insulation; and construction-related packaging. Construction and demolition waste does not include waste material containing friable asbestos; garbage; furniture and appliances from which ozone depleting chlorofluorocarbons have not been removed in accordance with the provisions of the federal Clean Air Act; electrical equipment containing hazardous materials; tires; drums; and containers, even though such wastes resulted from construction and demolition activities. Clean rubble that is mixed with other construction and demolition waste during demolition or transportation is normally considered to be construction and demolition waste.

Construction documents: The working drawings and the specifications prepared by the architect. The waste management requirements for a project are usually stated in the specifications.

Construction-related packaging: Small quantities of packaging wastes that are generated in the construction or renovation of structures.

Construction Specifications Institute (CSI) divisions of construction: A national association promoting consistency and professionalism in the writing of construction specifications. The organization publishes the CSI Divisions of Construction, which is widely used as the organizational format for master specifications and construction cost-estimating systems.

Container rental: Monthly fee for having a compactor or Dumpster on-site.

Contract for construction: The agreement between the owner and contracting entity (whether general contractor or construction manager at risk) for construction of the work represented in the construction documents. Also known as the construction contract, or the agreement between owner and contractor.

Copper: Definitions of product types for recyclers; #1 copper: 16-gauge or thicker (about pencil lead); free of paint, solder, fittings, plating. #2 copper: thinner than 16-gauge; can have some other metals, plated. Copper light (sheet): roofing, sheet, gutters, downspouts; no insulation or other materials.

Corrugated cardboard [also known as *old corrugated cardboard* (OCC) and/or cardboard box]: Paper product made of unbleached kraft fiber, with two heavy outer layers and a wavy inner layer to provide strength. Commonly used as a shipping container.

Crumb rubber: Rubber that has been grounded into small pieces.

Cullet: Clean, generally color-sorted crushed glass used to make glass products.

Deconstruction (also referred to as "soft demolition"): Deconstruction is often considered a sustainable alternative to conventional building demolition. In a deconstruction, hazardous materials are removed, reusable building materials are salvaged, demolition materials are recycled, and only a small portion of waste ends up in the landfill.

Deconstruction is also a specific phase of demolition in which materials suitable for reuse or recycling are removed from the building prior to general demolition.

Demofill: Landfill for construction and demolition waste only.

Demolition debris: Waste resulting from demolition operations on pavement, buildings, or other structures that includes lumber, drywall, concrete, pipe, brick, glass, electrical wire, and rubble.

Densification: The process of packing recyclables closely together, such as baling or re-rolling, to facilitate shipping and processing.

Diversion rate: A measure of the amount of waste being diverted from the municipal solid waste stream, either through recycling or composting.

Drywall: Drywall is an internal wall material made of gypsum, often covered on both sides with a paper facing. Drywall is also referred to as gypsum board, wallboard, plasterboard, and Sheetrock™.

Ebony: Metal scrap consisting of red brass scrap, valves, machinery bearings, and other machinery parts including miscellaneous castings made of copper, tin, zinc, and/or lead. Ebony must not include semi-red high-copper brass castings, railroad car boxes and other similar high-lead alloys, cocks and faucets, closed water meters, ingots and burned brass, aluminum, silicon and manganese bronzes, iron, and nonmetallics.

Eco-industrial park: A community of manufacturing and service businesses located together on a common property using each others' wastes as materials for production.

Electronic waste: Sometimes referred to as e-waste, this is a term applied to consumer and business electronic equipment that is no longer useful. E-waste includes computers, televisions, radios, CD players, and other electronic equipment.

Embodied energy: The sum total of the energy necessary—from raw material extraction, transport, manufacturing, assembly, and installation, plus the capital, environmental, and other costs—used to produce a service or product from its beginning through its disassembly, deconstruction, and/or decomposition.

Extremely hazardous waste: Any hazardous waste or mixture of hazardous wastes which, when exposed to humans, may result in death, disabling personal injury, or serious illness.

Ferrous metals: Metal containing iron (such as steel) in sufficient quantities to allow for magnetic separation.

Field superintendent: The contractor's day-to-day site manager. Responsible for coordination of subcontractor activities on the site, jobsite safety, and expediting progress of the work.

Forest residues: Fibrous by-products of harvesting, manufacturing, extractive, or woodcutting processes.

Friable asbestos: Asbestos incorporated into products such as floor tiles. It is in powder form or crumbles when broken or compressed, and may only be accepted at licensed landfills. Friable asbestos is not as hazardous as non-friable asbestos, but must be removed by protected personnel using safe procedures.

Gaylord: A 1.4-yd^3 cardboard container used to store loose materials.

General contractor: The prime or main contractor on a project, often responsible for coordinating and scheduling of other prime contractors. If there are multiple prime contracts on a project, the general contractor is the one responsible for general construction.

Generation data: Information on waste amounts derived from actual waste materials produced, usually determined by assessing waste bins on-site.

Granulator: A machine that produces small plastic particles.

Green building: The standard for construction that minimizes the effect of the built environment on the natural and social landscape.

Green Globes™: Green Globes is an environmental assessment, education, and rating system that is promoted in the United States by the Green Building Initiative, a Portland, Oregon-based nonprofit organization.

Greenwashing (also known as faux green): To falsely claim a product is environmentally sustainable.

Groundwater: The water contained in porous underground strata as a result of infiltration from the surface. Groundwater must be protected from waste storage runoff during construction.

Handler: A company that performs at least one of the following processes on collected recyclables: sorting, baling, shredding, or granulating.

Hauler: A company that provides transportation (hauling) services and vehicles from the construction site to a recycling center.

Hazardous waste: Waste that meets any of the following criteria: easily ignitable under ordinary temperature and pressure; readily supplies oxygen or reactive gas to a fire; is corrosive (highly acidic or caustic); is explosive or generates toxic gas; is acutely toxic to animals if it comes into contact with skin or is inhaled, eaten, or drunk; or contains toxic chemicals that can be dissolved in an acidic environment such as a landfill. Certain components of some electronic products contain materials that render them hazardous and unsuitable for recycling.

High-density polyethylene (HDPE): A plastic resin commonly used to make milk jugs, detergent containers, and base cups for plastic soda bottles. The standard plastic code for HDPE is #2.

High-grade waste paper: The most valuable waste paper for recycling. High-grade waste paper can be substituted for virgin wood pulp in making paper. Examples of high-grade waste paper include letterhead stationery and computer paper.

Hog fuel: A specific grade of ground up wood and bark. Hog fuel varies in size but is generally between ½ in and 6 in screen size. In some areas of the country, fuel from C&D recycling facilities is used to fuel boilers for wood and paper processing or other industries.

Hot-mix asphalt (HMA): Asphalt paving material produced by heating asphalt binder and mixing it with aggregate that has been dried to appropriate moisture content.

ICC-ES SAVE program: International Code Council *Sustainable Attributes Verification and Evaluation*™ (SAVE™) program. A program created to verify manufacturers' claims regarding the sustainable attributes of their products. Product evaluation under this program results in a Verification of Attributes Report™, which provides technically accurate product information that can be helpful to those seeking to qualify for points under major green rating systems.

Inappropriate for processing: Loads of C&D materials entering a facility of which less than 90 percent, as determined by the processor, can be sent on by the facility for recycling. Such loads are usually charged a higher rate.

Industrial scrap: Recyclables generated by manufacturing processes, such as trimmings and other leftover materials, or recyclable products that have been used by industry but are no longer needed, such as buckets, shipping containers, signs, pallets, and wraps.

Inert debris: Those materials that are virtually inert, such as rock, dirt, brick, concrete, or other rubble.

Insulation: Insulation comes in many different forms. Fiberglass is the most common. It consists of flexible fragments of spun or woven glass formed into batts, though it is sometimes also blown into cavities. Cellulose insulation is also very common and usually consists of shredded newspaper mixed with a binder and fire retardant. Another type of insulation, vermiculite insulation, is a pebble-like, pour-in product that is light brown or gold in color. Vermiculite insulation sometimes contains asbestos fibers.

International Code Council (ICC): ICC, a membership association dedicated to building safety, fire prevention, and energy efficiency, develops the codes used to construct residential and commercial buildings, including homes and schools. Most U.S. cities, counties, and states that adopt codes utilize ICC codes. ICC is working to develop the International Green Construction Code (IGCC) and the ICC-ES SAVE program.

International Green Construction Code (IGCC): A model green construction code developed by the International Code Council in collaboration with other national building-related organizations. The IGCC is coordinated with existing International Codes that span the spectrum of the industry from building, to energy conservation, and fire safety, plumbing, mechanical fuel gas, and existing buildings among others.

Junkyard: A lot or property where worn-out or discarded items, metal, or other scrap material is stored for possible resale.

Landfill (also know as a sanitary landfill): A specially engineered site for disposing of solid waste on land, constructed so that it will reduce hazards to public health and safety.

Lead: A metal found naturally in the environment as well as in manufactured products. The major sources of lead emissions have historically been motor vehicles and industrial sources. Lead may be encountered as part of the demolition process of a wide range of facilities.

Lead-based paint: Paint or other surface coatings that contain lead equal to or in excess of 1.0 mg/cm^2 or more than 0.5 percent by weight (5000 ppm).

LEED™: The acronym for Leadership in Energy and Environmental Design, a green building rating criteria developed by the U.S. Green Building Council. The LEED rating system is nationally recognized as the primary green building standard.

Life cycle: The stages of a product, beginning with the acquisition of raw materials, continuing with manufacture, construction, and use, and concluding with a variety of recovery, recycling, or waste management options.

Locally sourced materials: Materials obtained from within a defined radius around a project site, in order to support the local economy and to reduce transportation costs and energy.

Low-density polyethylene (LDPE): A plastic used in shopping bags and garbage bags. The standard plastic code for LDPE is #4.

Manual separation (or segregation): The sorting of recyclables from other waste by hand.

Material recovery facility (MRF): This is a general term used to describe a waste-sorting facility. Mechanical separation, separation by hand, or a combination of both procedures is used to recover recyclable materials from other waste.

Mercury: A toxic metal that can cause harm to people and animals, including nerve damage and birth defects. Mercury released into the environment can contaminate the air and enter streams, rivers, and the ocean, where it contaminates fish that people eat.

Mixed C&D waste: C&D materials containing both recyclable and non-recyclable C&D materials that have not been source separated. C&D waste is considered to be "mixed" C&D waste if it contains more than 10 percent—but less than 90 percent—recyclable C&D waste by volume. At a mixed C&D recycling facility, different recyclables are sorted from a load of mixed debris. A load of mixed C&D generally includes drywall, metal, untreated wood, yard trimmings, and small amounts of inert materials.

Mixed paper: Waste paper of various kinds and qualities. Examples include stationary, notepads, manila folders, and envelopes.

Nonferrous metals: Metals such as copper, brass, bronze, aluminum bronze, lead, pewter, zinc, and other metals to which a magnet will not adhere.

Non-friable asbestos: Asbestos not in powder form and does not crumble when broken or compressed in the hand. Considered hazardous and must be removed under controlled conditions by licensed and protected personnel.

Nonrenewable resources: Natural materials that are considered finite because of their scarcity, the long time required for their formation, or their rapid depletion.

Organic waste: Discarded living material such as vegetative and food waste.

Pallet: A wooden platform placed underneath large items so they may be picked up and moved by a forklift.

Paperboard: Heavyweight grades of paper commonly used for packaging products like cereal boxes. Paperboard is classified differently from corrugated cardboard for recycling purposes.

PCBs: Polychlorinated biphenyls are a class of industrial chemicals manufactured from 1930 to 1977 for use in electrical and hydraulic products. PCBs are still present in the environment because of their persistence and accumulation, and may be encountered in items removed from demolished buildings.

Plaster: A mixture of cement or gypsum plaster with sand, perlite, or vermiculite, and sometimes lime, added to form an interior wet wall system when applied to lath work or plasterboard. Stucco is an exterior form of plaster.

Plastic lumber: A lumber product made from recycled plastics or a composite of wood fiber and plastic. Resistant to water, chemicals, and pests, plastic lumber is suggested for decking and light construction; it is not suitable for structural framing.

Polyethylene terephthalate (PET): A plastic commonly used to make soft drink bottles and other food packaging like ketchup and salad dressing bottles. The standard plastic code for PET is #1.

Polypropylene (PP): Plastic material that is used to manufacture dairy tubs, lids, and straws. The standard plastic code for PP is #5.

Polystyrene (PS): A lightweight plastic material often used in food services. Polystyrene products include tray, plates, bowls, cups, and hinged containers. The standard plastic code for PS is #6.

Polyvinyl chloride (PVC): Plastic material used to manufacture piping and food and cosmetic containers. The standard plastic code for PVC is #3.

Postconsumer: Types of recycled-content products containing materials that have been previously used by consumers and then reprocessed into new products.

Postmanufacture content: Also known as postmanufacture waste, this term refers to waste that was created by a manufacturing process and is subsequently only used as a constituent in another manufacturing process. Similar to postindustrial waste, although that term is typically applied to waste products recycled within the same process.

Preconsumer (also known as postindustrial recycled-content): These products contain waste materials created as a result of manufacturing processes, that were then used in the manufacture of new products.

Postindustrial recycled content: Indicates that manufacturing waste has been cycled back into the production process. These products do not represent the significant resource savings that postconsumer products do, but are far preferable to those that use virgin materials.

Project manager: Typically, the employee of the contractor or construction manager responsible for the overall management of the contractor's operations on the site and fulfillment of contractual obligations with the owner. The architect's employee responsible for the professional services of the project may also be referred to as a project manager, as may an owner's representative responsible for representing the owner's interests on the site.

Pull fee (also known as a haul fee): The charge for collecting and transporting waste to a waste disposal facility.

Pure loads of recyclable C&D waste: Loads of single-type or mixed types of recyclable C&D waste that contain at least 90 percent recyclable C&D waste materials by volume.

Reclaimed lumber: Wood that has been removed from defunct structures or logs that have sunk in rivers during transport. It has all the advantages—hard, stable, free of knots—of old-growth timber, without the need for continued logging of already depleted forests.

Reclaimer (reprocessor): A company that performs at least one of the following processes on collected recyclables: washing/cleaning, pelletizing, or manufacturing a new product.

Recycle: The separation of construction waste or demolition materials into separate recycle categories for reuse into marketable materials. Examples of recycling include separating wood waste for recycling into paper pulp, or taking soil to a topsoil facility for reprocessing into topsoil.

Recyclables: Any materials that will be used or reused as an ingredient in an industrial process to make a product, or as an effective substitute for a commercial product. Common construction recyclables include paper, glass, concrete, plastic, steel, and asphalt.

Recycled asphalt pavement (RAP): Pavement reclaimed and reused in new pavement. Some states allow for use of RAP along with recycled asphalt shingles (RAS).

Recycled asphalt shingles (RAS): Properly ground-up shingle waste prepared for end use.

Recycled content: The amount of a product's or package's weight that is composed of materials that have been recovered from waste. Recycled content may include preconsumer and postconsumer materials.

Resource Conservation and Recovery Act (RCRA): RCRA was enacted by Congress in 1976 as an amendment to the 1965 Solid Waste Disposal Act. The goals of RCRA are to protect human health and the environment from the hazards posed by waste disposal, conserve energy and natural resources through waste recycling and recovery, reduce the amount of waste generated, and ensure that wastes are managed in a manner that is protective of human health and the environment.

Reuse: The use of demolished materials on the same site for the same or different purpose. Examples include grinding concrete or asphalt for reuse on-site, or reusing framing lumber and steel sections.

Rock boxes: Slang for the container holding large amounts of mixed inert materials, such as concrete, brick, and block masonry.

Salvage: The act of removing construction or demolition waste from an existing building to be reused on that site or elsewhere in the same form, or the products removed as a result of salvage. Examples of salvage include removing brick, tile, ornamental architectural items, lumber, doors, or plumbing fixtures.

Scrap: Discarded or rejected industrial waste material often suitable for recycling.

Scrap material processing industry: Any person or company who accepts, processes, and markets recyclables.

Segregation: The systematic separation of solid waste into designated categories for pickup by recyclers.

Shredder: A machine that tears or grinds material to reduce its size.

Solid Waste Disposal Act: A federal law passed in 1965 and amended in 1970 that addresses waste disposal methods, waste management, and resource recovery.

Solid waste processing facility: An incinerator, composting facility, household hazardous waste facility, waste-to-energy facility, transfer station, reclamation facility, or any other location where solid wastes are consolidated, temporarily stored, salvaged, or otherwise processed prior to being transported to a final disposal site. This does not include scrap metal material recycling and processing facilities.

Source reduction: Minimizing waste at the source of generation; preventing waste before it is generated.

Source separated: Segregating recycled material into separate containers for pickup by a recycler and transfer to a facility for processing. Source separation requires more site area and individual containers, but makes tracking and quality control of the materials easier. Source separation is normally used for materials such as drywall, concrete, carpet, plastics, and ceiling tiles.

Substitution: A product or process proposed by the contractor in lieu of the specified product or process, offered either with a credit for a lesser product or with no credit for an equivalent product. Typically, a substitution must be expressly approved by the architect prior to implementation or installation.

Tear-off shingles: Postconsumer shingle waste or construction shingle waste. Shingle waste removed from roofs, as well as scrap from new roofing projects.

Tin: In the recycling industry, tin is classified as thin steel of any size, but less than 1/8 in thick.

Tipping fee (also known as disposal fee): The price trash haulers pay to dispose of their waste at a landfill. The fee is usually dollars per ton.

Total inbound tons: The total tons of recyclable C&D waste, mixed C&D waste, and nonrecyclable C&D waste entering a receiving facility.

Transfer station: Any facility where solid wastes are transferred from one vehicle to another, or where solid wastes are stored and consolidated before being transported elsewhere.

Trash (garbage): Material considered worthless, unnecessary, or offensive that is usually thrown away; any product or material unable to be reused, returned, recycled, or salvaged.

Treated wood: Natural wood that has been treated with any of the following: creosote; oil-borne preservatives, including pentachlorophenol and copper naphthenate; water-borne preservatives, including chromated copper arsenate (CCA); or other chemicals or treatments that are hazardous to humans.

UBC: An acronym for a "used beverage container"; the term usually refers to plastic soda bottles and aluminum cans.

Universal wastes: Wastes that are hazardous upon disposal but pose a lower risk to people and the environment than do other hazardous wastes. State and federal regulations identify which unwanted products are universal wastes, and provide simple rules for handling and recycling them. Examples of universal wastes are televisions, computers, computer monitors, batteries, and fluorescent lamps.

Vinyl (V): A common type of plastic used to make shampoo bottles and other containers. The standard plastic code is #3.

Virgin resources: Resources using raw materials that have not been used before.

Volatile organic compounds (VOCs): Volatile organic compounds are a principal component in atmospheric reactions that form ozone and photochemical oxidants. VOCs are emitted from diverse sources, including automobiles, chemical manufacturing facilities, drycleaners, paint shops, and other commercial and residential sources that use solvent and paint.

Waste: Material that has been discarded because it has worn out, is used up, or is no longer needed.

Waste audit: An inventory of the amount and type of solid waste that is produced at a specific site as a result of construction or demolition activities.

Waste exchange: Two or more companies exchanging materials that would otherwise be discarded. The term also refers to an organization with electronic and/or catalog networks to match companies that want to exchange their materials.

Waste-to-energy facility: A facility that processes solid waste to produce energy or fuel.

Waste reduction: A construction site management strategy that encourages workers and subcontractors to generate less trash through practices such as reuse, recycling, and buying products with less packaging.

Waste reduction and recycling plan: A written plan for the recycling of project C&D debris.

Wetlands: A lowland area, such as a marsh or swamp, saturated with water. Wetlands are crucial wildlife habitats, and important for flood control and maintaining the health of surrounding ecosystems.

White goods: Appliances such as refrigerators, stoves, water heaters, washing machines, dryers, and air conditioners.

Yellow brass scrap: Mixed scrap metal consisting of yellow brass solids, including brass castings, rolled brass, rod brass, tubing, and miscellaneous plated brass. Must be free of manganese-bronze, aluminum-bronze, radiators or radiator parts, iron, and excessively dirty or corroded materials. Must also be free of any type of munitions, including bullet casings.

INDEX

International Green Construction Code (IGCC)

The International Code Council (ICC) published a public version of the *International Green Construction Code* (IGCC) in March of 2010. This Public Version (PV 1.0) serves as the base document in the code development process leading to the publication of the 2012 IGCC. Partnering in this effort were Cooperating Sponsors the American Institute of Architects (AIA) and ASTM International, which have supported the development of an adoptable and enforceable green building code.

The IGCC is also augmented with *ANSI/ASHRAE/USGBC/IES Standard 189.1, Standard for the Design of High Performance, Green Buildings Except Low-Rise Residential Buildings*, as an alternate path of compliance. The IGCC provides the building industry with language that both broadens and strengthens building codes in a way that will accelerate the construction of high-performance, green buildings. It provides a vehicle for jurisdictions to regulate green for the design and performance of new and renovated buildings in a manner that is integrated with existing codes, allowing communities to reap the rewards of improved design and construction practices.

The code includes both the technical content of the IGCC and Standard 189.1 addressing technical requirements such as water use efficiency, indoor environmental quality, energy efficiency, materials and resource use, and the building's impact on its site and its community.

For the latest code development activities related to the IGCC, interested individuals can visit the ICC website at www.iccsafe.org.

Sustainable Attributes Verification and Evaluation™ (SAVE™)

The number of new products and innovative approaches that are introduced into the building construction industry on a continuing basis has been phenomenal. A large number of these new products or practices claim to be "green" or consistent with and promoting the goals of sustainable construction. The question of sustainability becomes more complicated considering the global market and the fact that many products are sold internationally. Because it is difficult for designers, contractors or code officials to verify the credibility of each and every claim regarding sustainable attributes, it is important that there is a reliable source to verify such claims are legitimate. Verification requires a methodology and a standardized process by which to evaluate the degree of "greenness" and the sustainable attributes of construction materials, elements and assemblies that establish a building's green performance.

Because of the large number of innovative green products that are being introduced, ICC Evaluation Service (ICC-ES) has created a program to address the issue of green evaluation called the Sustainable Attributes Verification and Evaluation™ (SAVE™) Program.

The SAVE™ Program from ICC-ES® provides the most trusted third-party verification available today for sustainable construction products. Under this program, ICC-ES® evaluates a product's conformance to the requirements contained in current green standards. The SAVE™ Program may assist in identifying products that have been evaluated to multiple SAVE™ guidelines and multiple green building rating systems, standards and codes, such as US Green Building Council's LEED, Green Building Initiative's Green Globes, National Association of Home Builders and International Code Council's National Green Building Standard (ICC 700-2008) and the California Green Building Standards Code.

The ICC-ES SAVE Verification of Attributes Reports (VAR) are easily accessible online at www.icc-es.org/save. A sample VAR is shown.

ICC EVALUATION SERVICE

Most Widely Accepted and Trusted

ICC-ES SAVE Verification of Attributes Report™

VAR-1053

Issued July 1, 2009

This report is subject to re-examination in one year.

www.icc-es.org/save | 1-800-423-6587 | (562) 699-0543   *A Subsidiary of the International Code Council®*

DIVISION 07—THERMAL AND MOISTURE PROTECTION
Section 07 21 16—Building Insulation
Section 07210—Building Insulation

REPORT HOLDER:

A–1 Insulation, Inc.
123A Rocky Road
Asphalt, CA 43210
(123) 765-4321
www.a1insulation.com
jv@a1insulation.com

EVALUATION SUBJECT:

A–1 Insulation

1.0   EVALUATION SCOPE

Compliance with the following evaluation guideline:

ICC-ES Evaluation Guideline for Determination of Biobased Material Content (EG102), dated October 2008.

2.0   USES

A–1 Insulation is a semirigid, low-density, cellular isocyanate foam plastic insulation that is spray-applied as a nonstructural insulating component of floor/ceiling and wall assemblies.

3.0   DESCRIPTION

A–1 Insulation is a two component system with a nominal density of 1.0 pcf (16 kg/m³). The insulation is produced by combining the two components on-site. Water is used as the blowing agent and reacts with the isocyanate, which releases a gas, causing the mixture to expand. The mixture is spray-applied to the surfaces intended to be insulated.

The insulation contains the minimum percentage of biobased content as noted in Table 1.

4.0   CONDITIONS

Evaluation of A–1 Insulation for compliance with the International Codes is outside the scope of this evaluation report. Evidence of compliance must be submitted by the permit applicant to the Authority Having Jurisdiction for approval.

5.0   IDENTIFICATION

The A–1 Insulation spray foam insulation described in this report is identified by a stamp bearing the manufacturer's name and address, the product name, and the VAR number (VAR-1053).

TABLE 1 – BIOBASED MATERIAL CONTENT SUMMARY

% MEAN BIOBASED CONTENT	METHOD OF DETERMINATION
15% (+/–3%)[1]	ASTM D6866

[1]Based on precision and bias cited in ASTM D 6866.
